Iterative Aggregation Theory

PURE AND APPLIED MATHEMATICS

A Program of Monographs, Textbooks, and Lecture Notes

MONOGRAPHS AND TEXTBOOKS IN PURE AND APPLIED MATHEMATICS

1. *K. Yano,* Integral Formulas in Riemannian Geometry (1970) *(out of print)*
2. *S. Kobayashi,* Hyperbolic Manifolds and Holomorphic Mappings (1970) *(out of print)*
3. *V. S. Vladimirov,* Equations of Mathematical Physics (A. Jeffrey, editor; A. Littlewood, translator) (1970) *(out of print)*
4. *B. N. Pshenichnyi,* Necessary Conditions for an Extremum (L. Neustadt, translation editor; K. Makowski, translator) (1971)
5. *L. Narici, E. Beckenstein, and G. Bachman,* Functional Analysis and Valuation Theory (1971)
6. *D. S. Passman,* Infinite Group Rings (1971)
7. *L. Dornhoff,* Group Representation Theory (in two parts). Part A: Ordinary Representation Theory. Part B: Modular Representation Theory (1971, 1972)
8. *W. Boothby and G. L. Weiss (eds.),* Symmetric Spaces: Short Courses Presented at Washington University (1972)
9. *Y. Matsushima,* Differentiable Manifolds (E. T. Kobayashi, translator) (1972)
10. *L. E. Ward, Jr.,* Topology: An Outline for a First Course (1972) *(out of print)*
11. *A. Babakhanian,* Cohomological Methods in Group Theory (1972)
12. *R. Gilmer,* Multiplicative Ideal Theory (1972)
13. *J. Yeh,* Stochastic Processes and the Wiener Integral (1973) *(out of print)*
14. *J. Barros-Neto,* Introduction to the Theory of Distributions (1973) *(out of print)*
15. *R. Larsen,* Functional Analysis: An Introduction (1973) *(out of print)*
16. *K. Yano and S. Ishihara,* Tangent and Cotangent Bundles: Differential Geometry (1973) *(out of print)*
17. *C. Procesi,* Rings with Polynomial Identities (1973)
18. *R. Hermann,* Geometry, Physics, and Systems (1973)
19. *N. R. Wallach,* Harmonic Analysis on Homogeneous Spaces (1973) *(out of print)*
20. *J. Dieudonné,* Introduction to the Theory of Formal Groups (1973)
21. *I. Vaisman,* Cohomology and Differential Forms (1973)
22. *B.-Y. Chen,* Geometry of Submanifolds (1973)
23. *M. Marcus,* Finite Dimensional Multilinear Algebra (in two parts) (1973, 1975)
24. *R. Larsen,* Banach Algebras: An Introduction (1973)
25. *R. O. Kujala and A. L. Vitter (eds.),* Value Distribution Theory: Part A; Part B: Deficit and Bezout Estimates by Wilhelm Stoll (1973)
26. *K. B. Stolarsky,* Algebraic Numbers and Diophantine Approximation (1974)
27. *A. R. Magid,* The Separable Galois Theory of Commutative Rings (1974)
28. *B. R. McDonald,* Finite Rings with Identity (1974)
29. *J. Satake,* Linear Algebra (S. Koh, T. A. Akiba, and S. Ihara, translators) (1975)

30. *J. S. Golan,* Localization of Noncommutative Rings (1975)
31. *G. Klambauer,* Mathematical Analysis (1975)
32. *M. K. Agoston,* Algebraic Topology: A First Course (1976)
33. *K. R. Goodearl,* Ring Theory: Nonsingular Rings and Modules (1976)
34. *L. E. Mansfield,* Linear Algebra with Geometric Applications: Selected Topics (1976)
35. *N. J. Pullman,* Matrix Theory and Its Applications (1976)
36. *B. R. McDonald,* Geometric Algebra Over Local Rings (1976)
37. *C. W. Groetsch,* Generalized Inverses of Linear Operators: Representation and Approximation (1977)
38. *J. E. Kuczkowski and J. L. Gersting,* Abstract Algebra: A First Look (1977)
39. *C. O. Christenson and W. L. Voxman,* Aspects of Topology (1977)
40. *M. Nagata,* Field Theory (1977)
41. *R. L. Long,* Algebraic Number Theory (1977)
42. *W. F. Pfeffer,* Integrals and Measures (1977)
43. *R. L. Wheeden and A. Zygmund,* Measure and Integral: An Introduction to Real Analysis (1977)
44. *J. H. Curtiss,* Introduction to Functions of a Complex Variable (1978)
45. *K. Hrbacek and T. Jech,* Introduction to Set Theory (1978)
46. *W. S. Massey,* Homology and Cohomology Theory (1978)
47. *M. Marcus,* Introduction to Modern Algebra (1978)
48. *E. C. Young,* Vector and Tensor Analysis (1978)
49. *S. B. Nadler, Jr.,* Hyperspaces of Sets (1978)
50. *S. K. Segal,* Topics in Group Rings (1978)
51. *A. C. M. van Rooij,* Non-Archimedean Functional Analysis (1978)
54. *L. Corwin and R. Szczarba,* Calculus in Vector Spaces (1979)
53. *C. Sadosky,* Interpolation of Operators and Singular Integrals: An Introduction to Harmonic Analysis (1979)
54. *J. Cronin,* Differential Equations: Introduction and Quantitative Theory (1980)
55. *C. W. Groetsch,* Elements of Applicable Functional Analysis (1980)
56. *I. Vaisman,* Foundations of Three-Dimensional Euclidean Geometry (1980)
57. *H. I. Freedman,* Deterministic Mathematical Models in Population Ecology (1980)
58. *S. B. Chae,* Lebesgue Integration (1980)
59. *C. S. Rees, S. M. Shah, and C. V. Stanojević,* Theory and Applications of Fourier Analysis (1981)
60. *L. Nachbin,* Introduction to Functional Analysis: Banach Spaces and Differential Calculus (R. M. Aron, translator) (1981)
61. *G. Orzech and M. Orzech,* Plane Algebraic Curves: An Introduction Via Valuations (1981)
62. *R. Johnsonbaugh and W. E. Pfaffenberger,* Foundations of Mathematical Analysis (1981)
63. *W. L. Voxman and R. H. Goetschel,* Advanced Calculus: An Introduction to Modern Analysis (1981)
64. *L. J. Corwin and R. H. Szcarba,* Multivariable Calculus (1982)
65. *V. I. Istrătescu,* Introduction to Linear Operator Theory (1981)
66. *R. D. Järvinen,* Finite and Infinite Dimensional Linear Spaces: A Comparative Study in Algebraic and Analytic Settings (1981)

67. *J. K. Beem and P. E. Ehrlich,* Global Lorentzian Geometry (1981)
68. *D. L. Armacost,* The Structure of Locally Compact Abelian Groups (1981)
69. *J. W. Brewer and M. K. Smith, eds.,* Emmy Noether: A Tribute to Her Life and Work (1981)
70. *K. H. Kim,* Boolean Matrix Theory and Applications (1982)
71. *T. W. Wieting,* The Mathematical Theory of Chromatic Plane Ornaments (1982)
72. *D. B. Gauld,* Differential Topology: An Introduction (1982)
73. *R. L. Faber,* Foundations of Euclidean and Non-Euclidean Geometry (1983)
74. *M. Carmeli,* Statistical Theory and Random Matrices (1983)
75. *J. H. Carruth, J. A. Hildebrant, and R. J. Koch,* The Theory of Topological Semigroups (1983)
76. *R. L. Faber,* Differential Geometry and Relativity Theory: An Introduction (1983)
77. *S. Barnett,* Polynomials and Linear Control Systems (1983)
78. *G. Karpilovsky,* Commutative Group Algebras (1983)
79. *F. Van Oystaeyen and A. Verschoren,* Relative Invariants of Rings: The Commutative Theory (1983)
80. *I. Vaisman,* A First Course in Differential Geometry (1984)
81. *G. W. Swan,* Applications of Optimal Control Theory in Biomedicine (1984)
82. *T. Petrie and J. D. Randall,* Transformation Groups on Manifolds (1984)
83. *K. Goebel and S. Reich,* Uniform Convexity, Hyperbolic Geometry, and Nonexpansive Mappings (1984)
84. *T. Albu and C. Năstăsescu,* Relative Finiteness in Module Theory (1984)
85. *K. Hrbacek and T. Jech,* Introduction to Set Theory, Second Edition, Revised and Expanded (1984)
86. *F. Van Oystaeyen and A. Verschoren,* Relative Invariants of Rings: The Noncommutative Theory (1984)
87. *B. R. McDonald,* Linear Algebra Over Commutative Rings (1984)
88. *M. Namba,* Geometry of Projective Algebraic Curves (1984)
89. *G. F. Webb,* Theory of Nonlinear Age-Dependent Population Dynamics (1985)
90. *M. R. Bremner, R. V. Moody, and J. Patera,* Tables of Dominant Weight Multiplicities for Representations of Simple Lie Algebras (1985)
91. *A. E. Fekete,* Real Linear Algebra (1985)
92. *S. B. Chae,* Holomorphy and Calculus in Normed Spaces (1985)
93. *A. J. Jerri,* Introduction to Integral Equations with Applications (1985)
94. *G. Karpilovsky,* Projective Representations of Finite Groups (1985)
95. *L. Narici and E. Beckenstein,* Topological Vector Spaces (1985)
96. *J. Weeks,* The Shape of Space: How to Visualize Surfaces and Three-Dimensional Manifolds (1985)
97. *P. R. Gribik and K. O. Kortanek,* Extremal Methods of Operations Research (1985)
98. *J.-A. Chao and W. A. Woyczynski, eds.,* Probability Theory and Harmonic Analysis (1986)
99. *G. D. Crown, M. H. Fenrick, and R. J. Valenza,* Abstract Algebra (1986)
100. *J. H. Carruth, J. A. Hildebrant, and R. J. Koch,* The Theory of Topological Semigroups, Volume 2 (1986)

101. *R. S. Doran and V. A. Belfi,* Characterizations of C*-Algebras: The Gelfand-Naimark Theorems (1986)
102. *M. W. Jeter,* Mathematical Programming: An Introduction to Optimization (1986)
103. *M. Altman*, A Unified Theory of Nonlinear Operator and Evolution Equations with Applications: A New Approach to Nonlinear Partial Differential Equations (1986)
104. *A. Verschoren*, Relative Invariants of Sheaves (1987)
105. *R. A. Usmani,* Applied Linear Algebra (1987)
106. *P. Blass and J. Lang,* Zariski Surfaces and Differential Equations in Characteristic $p > 0$ (1987)
107. *J. A. Reneke, R. E. Fennell, and R. B. Minton.* Structured Hereditary Systems (1987)
108. *H. Busemann and B. B. Phadke*, Spaces with Distinguished Geodesics (1987)
109. *R. Harte*, Invertibility and Singularity for Bounded Linear Operators (1987)
110. *G. S. Ladde, V. Lakshmikantham, and B. G. Zhang*, Oscillation Theory of Differential Equations with Deviating Arguments (1987)
111. *L. Dudkin, I. Rabinovich, and I. Vakhutinsky*, Iterative Aggregation Theory: Mathematical Methods of Coordinating Detailed and Aggregate Problems in Large Control Systems (1987)
112. *T. Okubo*, Differential Geometry (1987)
113. *D. L. Stancl and M. L. Stancl*, Real Analysis with Point-Set Topology (1987)
114. *H. Strade and R. Farnsteiner*, Modular Lie Algebras and Their Representations (1987)
115. *T. C. Gard*, Stochastic Differential Equations (1987)

Other Volumes in Preparation

Iterative Aggregation Theory

Mathematical Methods of Coordinating Detailed and Aggregate Problems in Large Control Systems

Lev M. Dudkin
Harvard University
Cambridge, Massachusetts

Ilya Rabinovich
Academy of Sciences of the U.S.S.R.
Kizil
Union of Soviet Socialist Republics

Ilya Vakhutinsky
All Union Scientific Research Institute of Oil
Moscow
Union of Soviet Socialist Republics

MARCEL DEKKER, INC. New York and Basel

Library of Congress Cataloging-in-Publication Data

Dudkin, L. M. (Lev Mikhailovich), [date].
Iterative aggregation theory.

(Monographs and textbooks in pure and applied mathematics ; v. 111)
Bibliography: p.
Includes index.
1. Iterative methods (Mathematics) 2. Economics--Statistical methods. I. Rabinovich, I. N. (Ilya Naumovich) II. Vakhutinsky, I. Ya. (Ilya Yakovlevich). III. Title.
QA297.8.D83 1987 519.4 87-13341
ISBN 0-8247-7570-8

MARCEL DEKKER, INC.
270 Madison Avenue, New York, New York 10016

Current printing (last digit):
10 9 8 7 6 5 4 3 2 1

PRINTED IN THE UNITED STATES OF AMERICA

Foreword

What is the essential core of the theory of central planning - that area of economic theory which encompasses the optimal use of resources in large, planned, nonmarket organizations? I suppose it must be the theory of optimization and decomposition, understood in a very broad sense. The "best coordination" of subunits, which constitutes the essence of economic planning, is really a theory of "iterative aggregation." This is because iterative aggregation theory deals with how to plan over time in large organizations where the more highly aggregated units understand and can communicate only in a more highly aggregated language, while the more disaggregated units understand and can communicate only in a more disaggregated language.

Iterative aggregation theory, then, is close to being a basic description of economic planning itself. That is why the present book is so interesting and important. In it, the authors, who played a key role in developing new algorithms for different planning models in the U.S.S.R., have constructed a general mathematical theory of iterative aggregation methods. They show clearly the connection between the new methods and other classical, decomposition, and iterative optimizing algorithms. The book demonstrates a sound mathematical basis for coordinating solutions of aggregate models of the higher levels of the planning system with solutions of the detailed models of the lower levels of the planning system.

Two basic consequences are developed. On the more purely mathematical side, these new methods appear to have essentially more rapid convergence properties than classical iterative, decomposition, or other older methods. That is why the book will be of interest to mathematicians.

On the economic side, the new methods have interesting economic interpretation as a way of ensuring coordination of local planning decisions with the aggregate calculations of the higher levels of the planning system.

This book will also be of interest to anyone concerned with economic planning or the coordination of large organizations. It deserves a broad audience. I hope it will be widely read.

Martin L. Weitzman
Professor of Economics
Massachusetts Institute of Technology
Cambridge, Massachusetts

Preface

Iterative aggregation methods are a new class of iterative mathematical methods. They appeared as a result of research into the problem of rigorously coordinating aggregated calculations at higher levels of planning systems with detailed calculations at lower levels of the systems. As with any new class of mathematical methods, they can be applied efficiently in diverse sciences and fields of human experience. This particular class of methods can be used especially well to describe the functioning of existing or projected hierarchical systems, where the task of coordinating problems with different levels of aggregation of information appears.

Our earlier books on iterative aggregation[1] were published in Russian. In these books we constructed iterative aggregation processes only for the solution of the specific system of equations of the interproduct input-output model, for the solution of the general linear programming model, and for some specific nonlinear optimization problems. In this book we also consider iterative aggregation processes for the solution of the systems of equations of general type and for the solution of unconstrained and constrained optimization problems of sufficiently general type.

In earlier books the emphasis was on the economic essence of large-scale and complex planning problems and also on the improvement of real-world planning processes with the help of iterative aggregation. In contrast, in this book we provide for the first time a comparative theoretical analysis of the various types of iterative aggregation algorithms with a view to determining their place among other computational methods. Our analysis has shown a significant connection between the ideas and methods of iterative aggregation and many of the methods of computational mathematics. We have found it possible to treat many well-known solution procedures for unconstrained optimization problems as limiting cases of iterative aggregation algorithms, and this was helpful in obtaining complete mathematical proofs of convergence for certain of the iterative aggregation algorithms (however, for some of the processes such proofs have been obtained only for limiting cases[2]). Our analysis has also made clear why iterative aggregation is one of the most effective general methods for solving both constrained and unconstrained optimization problems, regardless of the dimensions of the problem (in numerical experiments iterative aggregation algorithms have proven to be more effective than related classical iterative methods[3]).

From the very beginning the principal aim of iterative aggregation methods was to solve large-scale and complex planning problems by decomposing the initial problem into many subproblems and iteratively coordinating their solutions using an aggregate model. Each of the subproblems has substantially fewer dimensions and is much less complex than the initial one. Also, the coordinating aggregate model has few dimensions and is of simple structure.

Various iterative aggregation algorithms can be used for the solution of a given large-scale problem. In addition, different realizations of any given algorithm can be proposed: We can decompose the initial problem into different numbers and into different types of subproblems, we can use different mathematical methods for solving the subproblems and the aggregate problem, and we can allocate calculations and organize exchange of information between subproblems in different manners. This allows us, in a given hierarchical management system, to search for the most effective method and organization of calculations and exchange of information implementing our iterative aggregation procedure. The foregoing questions should also be considered as a part of iterative aggregation theory, and we take them into consideration in this book.

As is usual with linear programming methods, we use language suggested by applications to economics. However, we interpret iterative aggregation algorithms here as possible planning processes for a national economy as a whole, although at present iterative aggregation can be applied with benefit only in simpler hierarchical systems (in market economies: in very large firms and in subindustries regulated by the state; and in planned economies: in subindustries[4] and industries). With such an "impractical" interpretation we have in mind another significant role for these new methods—their use in the development of modern economic science.

Iterative aggregation theory changes some prevalent stereotypes of economic thinking. Indeed, at the first stages of research into the problem of coordination of aggregated and detailed planned calculations it became clear that for a modern developed economy such coordination is possible only by the free interchange of information between higher and lower levels of management without any "directives." In iterative aggregation processes for optimal production planning models, at every iteration the lower-level management boards will choose plans (more exactly, projects of plans) that differ from the blueprints just received from the higher-level management board if doing so benefits them[5] (if the lower-level management boards are forced to take the upper-level blueprints as obligatory, the processes of coordinating aggregate and detailed calculations will not converge at all). However, the aggregate plan is not a directive if it is not obligatory to carry it out.

Furthermore, it is unnecessary and useless to consider the final result of the procedure as a directive, as there is no reason for lower-level managements to deviate from the solution which they get at the end of the iterative aggregation process. Indeed, the upper-level management solution we get at the last iteration corresponds to the most profitable possible for lower-level management's differing choices.

Related to the fact of the nondirective nature of iterative aggregation planning is the possibility of constructing one-level iterative aggregation processes. In these processes, at every iteration every "lower-level management board" solves a combined problem in which its own detailed problem is combined with the aggregate problem of the entire economy. However, even in two- and multilevel processes, the higher-level management board is not really a directive center but only a regulative helper in the coordination of the individual programs of the national economy's subsystems.

In other words, the attempt to find a rigorous solution to the centralized state planning problem of avoiding imbalances between the supply and demand of specific products has led to an understanding that to achieve equilibrium in the economy, we must depart from the principles of centralized directive state planning.

These unexpected results lead us to a very interesting question. If we try to improve the processes of the market economy to escape from the unbalancing tendencies of the economy on the macro level, what results will we achieve? It is clear that this attempt will not lead to the construction of centralized directive state planning processes, but to what conclusions will such attempts lead us? It may be that the result will be just as unexpected and may be similar: The rational organization of a modern developed economy is that of an economy of intercoordinated programs of manufacture of products and services among the managers of various enterprises or subindustries, with the help of some regulatory boards that have only financial and penalty functions.

The acquaintance of Western economists, mathematicians, and system analysts with iterative aggregation methods can help to clear up this question and to contribute to the resolution of one of the world's main ideological conflicts.

The manuscript was translated from the Russian by Gregory Katzenelinbogen and the translation was edited by Victoria Poupko and Boris Youssin. David Bernstein of Northeastern University contributed to some sections and appendixes.

The authorship of the various parts is as follows: Introduction and Sections 3.1, 3.6, 3.7, 4.5, and 4.6, L. Dudkin; Sections 1.1 and 3.4, L. Dudkin and D. Bernstein; Chapters 2 and 5, and Sections 1.2, 1.3, 6.4, and 6.5, L. Dudkin, I. Rabinovich, and I.

Vakhutinsky; Sections 3.5, 4.1-4.4, 6.1, 6.2, and 6.6, L. Dudkin and I. Rabinovich; Sections 3.2, 3.3, and 6.3, L. Dudkin and I. Vakhutinsky; Appendixes 3.1 and 3.2, D. Bernstein and I. Rabinovich; Appendixes 3.3, 3.4, 4.1-4.3, 5.1, and 6.1, I. Rabinovich; Appendix 5.2, I. Rabinovich and I. Vakhutinsky; and Appendix 6.2, I. Vakhutinsky.

I wish to thank Professor Aron Katsenelinbogen for his help in arranging for the translation and publication of this book. I also thank Tamara Dudkin for her help in typing and retyping both the Russian and English versions.

Lev M. Dudkin

NOTES

[1] See Dudkin (1966, 1972), and Dudkin, ed. (1979). Main papers in English: Vakhutinsky, Dudkin, and Makarov (1973); Vakhutinsky et al. (1974, 1976); Vakhutinsky, Dudkin, and Shchennikov (1973); Dudkin and Ivankov (1975); Kasparson (1977); Dudkin and Kasparson (1979, 1983); Vakhutinsky, Dudkin, and Ryvkin (1979); Kuboniwa (1979, 1984); Miranker and Pan (1980); Chatelin and Miranker (1982) 1984); Shpalensky (1981); Mendelssohn (1982).

[2] We must note that for many classical iterative processes, proofs of convergence have also been obtained only for limiting cases.

[3] See Archangelsky et al. (1975), Dudkin, ed. (1979, Chaps. 7 and 8), Miranker and Pan (1980), Kuboniva (1984), and also Chapter 4 of this book. We designate all the algorithms that do not use aggregation as "classical."

[4] These methods have already been used for the solution of optimal planning problems in subindustries in the USSR (see Dudkin, ed., 1979, Chaps. 7 and 8).

[5] It is true that at every iteration the lower-level management boards are penalized for deviations from the master solution. However, if the gain due to the deviation exceeds the penalty, the management boards of subsystems deviate from the master solution.

Contents

Iterative Aggregation Theory

Introduction

Aggregate models are used when it is impossible or too difficult to solve large-scale mathematical problems. However, a one-time solution of an aggregate model does not really help in solving the initial problem. Indeed, when we construct an aggregate model, we create aggregated "known" indices of this model using information that is unknown in the initial model.

Let us illustrate this fact using the example of the simplest aggregate interindustry input-output model:

$$X_i = \sum_{j=1}^{J} a_{ij}X_j + b_i, \qquad i = 1, \ldots, J,$$
$$a_{ij} \geq 0, \qquad b_i \geq 0, \qquad \Sigma_i\, a_{ij} < 1, \tag{1}$$

where i and j are subscripts denoting types of industries, b_i is the known final demand for the products of the ith industry, a_{ij} is a known coefficient denoting the input of the products of the ith industry for manufacturing a unit output of the jth industry, and X_i is an unknown output of the ith industry.

This model is a result of the aggregation of a more concrete interproduct input-output model:

$$x_g = \sum_{q=1}^{G} \hat{a}_{gq}x_q + \hat{b}_g, \qquad g = 1, \ldots, G,$$
$$\hat{a}_{gq} \geq 0, \qquad \hat{b}_g \geq 0, \qquad \Sigma_g\, \hat{a}_{gq} < 1, \tag{2}$$

where g and q are subscripts denoting types of specific products, $\hat{b}_g$ is the known final demand for the gth product, $\hat{a}_{gq}$ is a known coefficient denoting the input of the gth product for manufacturing 1 unit of the qth product[1], and x_g is an unknown output of the gth product.

The interproduct input-output model differs from the interindustry input-output model in that the former deals with the production and allocation of specific products, whereas the latter deals with the production and allocation of industry outputs. The only formal difference between these two models is the difference in dimension. Realistically speaking, $J \leq G$.

These two systems of equations are connected by the following relations:

$$a_{ij} = \frac{\Sigma_{g \in G_i} \Sigma_{q \in G_j} \hat{a}_{gq} x_q}{\Sigma_{q \in G_j} x_q}, \qquad i, j = 1, \ldots, J;$$

$$b_i = \Sigma_{g \in G_i} \hat{b}_g, \qquad i = 1, \ldots, J,$$

where G_i is the set of products of the ith industry.

It is easy to see that to calculate the coefficients a_{ij} for the interindustry model (1) we use not only the known coefficients of the initial interproduct model (2), but also the unknown outputs of the initial model. This means that we cannot really solve the aggregate model before solution of the initial model.

In practice, economists construct the coefficients a_{ij} using approximations x_q^0 of the unknowns x_q. Using these approximations, one can calculate

$$a_{ij}^1 = \frac{\Sigma_{g \in G_i} \Sigma_{q \in G_j} \hat{a}_{gq} x_q^0}{\Sigma_{q \in G_j} x_q^0} \tag{3}$$

and then solve the interindustry model:

---

[1]The condition $\Sigma_g \hat{a}_{gq} < 1$ is specific only to the input-output model with products, measured in fixed prices (see Section 3.1).

$$X_i = \Sigma_j a^1_{ij} X_j + b_i. \tag{4}$$

Since the coefficients a_{ij} are only approximate values, a solution of system (4) does not necessarily represent a real solution of the interindustry model. We obtain such a solution only if in (3) we take $x^0_q = x^*_q$, where $\{x^*_q, q = 1, \ldots, G\}$ is an exact solution of the interproduct model. However, we do not know x^*_q until we solve the initial problem (2). That is why the solutions of interindustry input-output models in practice are too approximate to have any real value.

The so-called classical economic theory of aggregation studies the conditions of accurate aggregation and the problem of choosing the sets G_i in such a way as to minimize the errors of aggregation:

$$|X_i - \Sigma_{g \in G_i} x_g|, \qquad i = 1, \ldots, J.$$

The methods of this theory[1] are of great theoretical interest. However, they do not ensure escaping from or required minimization of errors. That is, the solutions X_i of system (4) are not necessarily close enough to $X^*_i = \Sigma_{g \in G_i} x^*_g$.

Moreover, even if we have $X_i = X^*_i$, it does not help us to solve the initial model at once. Indeed, if we add the linear equations

$$\Sigma_{g \in G_i} x_g = X^*_i, \qquad i = 1, \ldots, J$$

to the initial system (2), we will get G + J linear equations in G unknowns. Although when $X_i = X^*_i$, the additional J equations are linearly dependent on the original G equations, we still face the formidable problem of solving system (2).

Some works on classical aggregation also consider the problem of disaggregation, that is, the problem of getting disaggregated

---

[1] See, for example, Leontief (1947a,b); Hatanaka (1952); Malinvaud (1954); Teil (1957); Fisher (1958, 1965, 1968); Ara (1959); Yamada (1961); Dyumin and Arhangelsky (1966); Kossov (1966); Neudecker (1970); Morimoto (1970); and Dudkin, Khomyakhov, and Shchennikov (1973).

variables using the solution of the aggregate model. To obtain disaggregated variables, they use the following fixed-weight disaggregation operator:

$$\tilde{x}_g^1 = X_i^1 \frac{x_g^0}{\Sigma_{q \in G_i} x_q^0}, \qquad g \in G_i. \tag{5}$$

It is easy to see that the operator (5) preserves the same proportions between the values of variables $x_g = x_g^0$ of any of the separate subsets G_i, which were initially used to form the aggregate quantities X_i. If we use $\tilde{x}_g^1$ as new approximations for the calculation of new coefficients, we get

$$a_{ij}^1 = \frac{\Sigma_{g \in G_i} \Sigma_{q \in G_j} \hat{a}_{gq} \tilde{x}_q^1}{\Sigma_{q \in G_j} \tilde{x}_q^1} = \frac{\Sigma_{g \in G_i} \Sigma_{q \in G_j} \hat{a}_{gq} (x_q^0 / \Sigma_{g \in G_j} x_q^0)}{\Sigma_{q \in G_j} (x_q^0 / \Sigma_{g \in G_i} x_g^0)}$$

$$= \frac{\Sigma_{g \in G_i} \Sigma_{q \in G_j} \hat{a}_{gq} x_q^0}{\Sigma_{q \in G_j} x_q^0} = a_{ij}^0.$$

As $a_{ij}^1 = a_{ij}^0$, it does not make much sense to use a_{ij}^1 to obtain a new solution to the interindustry input-output model. Thus the interindustry model that uses the vector $\tilde{x}^1 = \{\tilde{x}_q^1, g = 1, \ldots, G\}$ gives the same solution as the interindustry model that uses the vector $x^0 = \{x_g^0, g = 1, \ldots, G\}$. We will get the same result if we calculate $\tilde{x}^2, \ldots, \tilde{x}^\sigma, \ldots$ in the same manner.

This created a deadlock in the use of disaggregation procedures in classical aggregation theory. This deadlock was broken by the iterative aggregation algorithm, proposed by Dudkin and Ershov (1965). The authors proposed considering $\tilde{x}_1$ not as a new approximation to the variables x_g, but as a stage for getting a new approximation.

In the Dudkin-Ershov algorithm one calculates the new approximation as the result of performing one iteration of the method of simple successive approximations over the $\tilde{x}^1$:

$$x_g^1 = \sum_{q=1}^{G} \hat{a}_{gq}\tilde{x}_q^1 + \hat{b}_g, \qquad g = 1, \ldots, G.$$

One iteration, σ, for the solution of the interproduct input-output model (2) in the Dudkin-Ershov algorithm can be described as follows:[1]

1. Compute the aggregate coefficients:

$$a_{ij}^{\sigma} = \frac{\Sigma_{g\in G_i} \Sigma_{q\in G_j} \hat{a}_{gq} x_q^{\sigma-1}}{\Sigma_{q\in G_j} x_q^{\sigma-1}}, \qquad i, j = 1, \ldots, J, \tag{6}$$

   where $G_i \cap G_j = \emptyset$ if $i \neq j$; $\cup_{i=1}^{J} G_i = \{1, \ldots, G\}$.

2. Solve the interindustry input-output model:

$$X_i = \Sigma_j a_{ij}^{\sigma} X_j + b_i, \qquad b_i = \Sigma_{g\in G_i} \hat{b}_g, \qquad i = 1, \ldots, J \tag{7}$$

   and denote the solution by X_i^{σ}.

3. Calculate the intermediate approximations:

$$\tilde{x}_g^{\sigma} = X_i^{\sigma} \frac{x_g^{\sigma-1}}{\Sigma_{q\in G_i} x_q^{\sigma-1}}. \tag{8}$$

4. Calculate the final approximations:

$$x_g^{\sigma} = \Sigma_q \hat{a}_{gq} \tilde{x}_q^{\sigma} + \hat{b}_g. \tag{9}$$

It is easy to see that only stage 2 of the iteration σ consists of a problem that must be solved in one place (the "center"). Other calculations could be made separately and simultaneously in parallel

---

[1] In Dudkin and Ershov (1965) the algorithm was described in another form, which we repeat exactly at the beginning of Chapter 3. Here we describe the algorithm in a form that can help us to understand the difference between and interrelationship of classical and iterative aggregation approaches.

in different places. To perform the entire procedure, however, the center must exchange certain information with the individual industries, and the industries must exchange certain information between themselves. Possible forms of the organization of the exchange of information have been studied by Vakhutinsky et al. (1976) and are discussed in Chapter 3 of the present book.

The main idea behind this and other two-level iterative aggregation algorithms as applied to planning processes (of which the interproduct input-output model is a simple case) is that we try to calculate first the industry outputs, then the intraindustry proportions, then again the industry outputs, and so on, until the plan is sufficiently in equilibrium.

We may assume that it is this very idea that underlies the practice of central planning. Indeed, it is easy to see that the procedure described above for coordinating detailed and aggregate calculations is closely analogous to the real process of drafting a balanced national economic plan. We can describe the process of drawing up a balanced plan in a planned economy in the following manner. At the top interindustry level, the central planning board determines outputs for the major industries, without giving a detailed breakdown of specific products. At the lower levels (industry, subindustry, etc.) of control the local planning boards solve local problems that determine the production of specific products in each individual industry, subindustry, and so on.

To draw up an interindustry plan, it is necessary to know the average aggregate input coefficients. Such a coefficient reflects the use of one industry's output in another industry's output. These coefficients cannot be calculated exactly until a detailed plan has been drawn up. However, if such a detailed plan were to exist, there would be no need to calculate the average input coefficients nor to draft the interindustry plan.

In practice, average coefficients (or equivalently, the weights given to the specific products) are computed on the basis of extrapolation and are substantiated by expert opinion. Another method of calculating average coefficients is to compute them with the help of formulas (3). These formulas are also the basis of the first method of getting average coefficients.

The industries' outputs are then calculated using those coefficients. The values of the outputs are passed on to the industries together with a number of other constraints (e.g., on capital investments). Within this framework the industries try to coordinate their plans at the more detailed level.

Although successive adjustments are incorporated into the planning processes of a national economy, these adjustment processes are not rigorous iterative processes for coordinating the solution of the interindustry problem with the solutions obtained from each of the industry problems. Only iterative aggregation has provided methods to ensure such rigorous coordination.

Since the construction of the first iterative aggregation algorithm, a large number of alternative iterative aggregation algorithms have been developed, not only for systems of input-output equations but also for systems of equations of general type and for solving unconstrained and constrained optimization problems. Each of these algorithms employs two stages at each iteration. The first stage, which we shall refer to as "aggregation transformation," involves the construction and solution of a fixed-weight aggregate model and fixed-weight disaggregation; the second stage entails performing one iteration of some classical[1] iterative process (which may use decomposition) over the disaggregated values of the variables.

The general form of classical algorithms can be described formally. Let the operator F, transforming x into $F(x)$, generate one iteration of some appropriate classical iterative process to be used in the second stage of the iterative aggregation algorithm. The iteration of the classical process is given by the sequence

$$x, F(x), F(F(x)), \dots \tag{10}$$

Instead of sequence (10), iterative aggregation algorithms usually apply the sequence of operators

[1]Conventionally, all the iterative processes that do not use aggregation will be referred to as "classical."

$$x,\ S(x),\ F(S(x)),\ S(F(S(x))),\ F(S(F(S(x)))),\ \ldots \tag{11}$$

where S(x) is the operator corresponding to the first (aggregation-disaggregation) stage of the iterative aggregation process.

For the first iterative aggregation algorithm the relevant operators to find the solution to the system of equations

$$x = Ax + b, \qquad A = (\hat{a}_{gq}) \geq 0, \qquad b = (\hat{b}_g) \geq 0, \qquad \Sigma_g \hat{a}_{gq} < 1$$

are the following: F transforms $\tilde{x}^\sigma$ into $x^\sigma = A\tilde{x}^\sigma + b$ and S transforms $x^{\sigma-1}$ into

$$\tilde{x}^\sigma = \left\{ X_i^\sigma \frac{x_g^{\sigma-1}}{\Sigma_{q\in G_i} x_q^{\sigma-1}},\ g \in G_i,\ i = 1, \ldots, J \right\},$$

where X_i^σ is the solution of the interindustry input-output equations whose coefficients are constructed on the basis of $x^{\sigma-1}$. In other words, operator S includes stages (6)-(8), and F includes stage (9).

The first algorithm can be interpreted as correcting a plan resulting from successive approximations using the solution of the aggregate input-output system (7). The essence of this correction is that the values of the variables are found that satisfy the interindustry solution in which the relative values of the variables within the outputs of industries are fixed, using the solution of the previous iteration.

Subsequently, this approach has led to the construction of a large number of other iterative aggregation algorithms for solving various problems. In each of these algorithms some classical iterative algorithm (10) is combined with some aggregation-disaggregation operator S. As an example, to solve the interproduct input-output model, the correction S given by (6)-(8) has been integrated with the Gauss-Seidel algorithm, Nekrasov algorithm, and others [see Dudkin and Kasparson (1979) and Section 3.6 of the present book].

However, for the solution of problems more complicated than the interproduct input-output model (2) with fixed prices, using the ideas of iterative aggregation requires more than a direct application of stages (6)-(9). Thus, in solving an interproduct input-output model with products measured in physical units,[1] it is necessary to modify somewhat the aggregation-disaggregation operator S used for solving problem (2) [see Dudkin and Rabinovich (1976) and Section 3.4 of the present book].

Special operators F and/or S were used to solve some optimal planning models in which iterative aggregation was applied only for a square input-output submatrix of the initial problem. [See Dudkin (1966, Chap. IV; 1972, Chap. V); Vakhutinsky, Dudkin, and Shchennikov (1973); and Section 4.4 of the present book.]

New types of iterative aggregation processes proposed for solving input-output equations [see Dudkin and Rabinivich (1978) and Section 3.5 of the present book] and algorithms for solving a system of linear equations of general type (see Chapter 4) and unconstrained optimization problems (Chapter 5) also use specific operators F and S.

In some of the algorithms for optimal planning models in which iterative aggregation was used for a square input-output submatrix, and in the algorithms for solving linear programming problems (see, e.g., Vakhutinsky and Dudkin, 1972; Vakhutinsky, Dudkin, and Ryvkin, 1979; and Dudkin, ed., 1979, the operator F includes the calculation of the dual solution (i.e., dual prices). These dual prices are used in the formulas of aggregation (i.e., in operator S and in operator F at the next iteration). We can see the same approach in the iterative aggregation algorithms for solving general constrained extremum problems, which are introduced in Chapter 6.

Such recent work as Miranker and Pan (1980), Babadshzanyan (1980), Shpalensky (1981), Chatelin and Miranker (1982), and Dudkin and Kasparson (1983) represents primarily new modifications of the aggregated problems (i.e., some new operators S).

Although iterative aggregation algorithms have been proposed for solving problems of different types and various algorithms have

---

[1] That is, when condition $\Sigma_g \hat{a}_{gq} < 1$ is not fulfilled and is replaced by the weaker condition of so-called productivity (see Section 3.1).

been proposed for each type of problem, all of them apply the sequence (11) with operators S and F. This is why we can consider the application of this sequence (11) as the main common feature of iterative aggregation processes.

In this book we consider some specific classes of iterative aggregation algorithms that use gradient methods. In these algorithms at every iteration we get, simultaneously, a gradient, which includes an intermediate approximation $\tilde{x}$, and the final approximation F(x). We can interpret these algorithms as some specific type, or modification, of scheme (11).

Since iterative aggregation processes are based on classical iterative processes, it is natural that Part I review the essentials of contemporary iterative methods, including methods that use decomposition. In Parts II and III we consider the main iterative aggregation methods for solving systems of equations and unconstrained and constrained optimization problems. These methods are compared both among themselves and with their classical precursors, with a view to finding the most effective solution processes or the most effective combination of solution processes.

PART I

REVIEW OF CONTEMPORARY METHODS OF MATHEMATICAL PROGRAMMING

1

General Methods for Solving Extremum Problems and Systems of Equations

1.1 SOME NOTIONS OF THE GENERAL THEORY OF EXTREMUM PROBLEMS AND SYSTEMS OF EQUATIONS

1.1.1 Extremal problems and systems of equations are usually formulated for an arbitrary number n of variables x_i, $i = 1, \ldots, n$. In geometric terms it means that we usually deal with an n-dimensional real vector space $\mathbb{R}^n = \{(x_1,\ldots,x_n) | x_i \in \mathbb{R}\}$, where $\mathbb{R}$ is the set of real numbers. Sometimes it is advantageous to allow x_n to have complex values, and that leads to dealing with an n-dimensional complex vector space $\mathbb{C}^n = \{(z_1,\ldots,z_n) | z_i \in \mathbb{C}\}$, where $\mathbb{C}$ is the set of complex numbers.

1.1.2 It is convenient to introduce some geometric notions for multidimensional vector spaces. The first of them is the Euclidean norm. The Euclidean norm (or the length) $\|x\|$ of a vector $x = (x_1, \ldots,x_n)$ is defined by $\|x\| = \sqrt{|x_1|^2 + \cdots + |x_n|^2}$, where x may be a real or a complex vector. A motivation for such a definition is provided by the Pythagorean theorem. Indeed, in two-dimensional space if $x = (x_1,x_2)$ is a vector whose length (hypotenuse) equals ℓ, then $\ell^2 = x_1^2 + x_2^2$. Therefore, $\|x\| = \ell = \sqrt{x_1^2 + x_2^2}$.

This norm has the following properties:

(a) $\|x + y\| \le \|x\| + \|y\|$ for any $x, y \in \mathbb{R}^n$ (or $\mathbb{C}^n$); this inequality is known as the triangle inequality.

(b) $\|kx\| = |k|\|x\|$ for any vector x and any number k.

(c) $\|x\| > 0$ for any nonzero vector x.

Often it is useful to consider other norms. We say that a real-valued function $\|\cdot\|$ defined on $\mathbb{R}^n$ or $\mathbb{C}^n$ that is a function assigning to any vector x a number $\|x\|$ is a norm if it satisfies conditions (a)-(c). In addition to the Euclidean norm (sometimes called the ℓ_2 norm) there is another widely used norm called the ℓ_1 norm. It is defined by $\|x\| = \Sigma_i |x_i|$.

1.1.3 Consider an arbitrary n × n matrix $A = (a_{ij})$, $1 \le i,j \le n$, with real coefficients. It defines the linear transformation T of $\mathbb{R}^n$ given by $Tx = z$, where $z = (z_1,\ldots,z_n)$, $z_i = \Sigma_j a_{ij}x_j$. This transformation has the following properties:

(a) $T(x + y) = Tx + Ty$ for any vectors x, y.

(b) $T(kx) = kTx$ for any vector x and any number k.

Conversely, any transformation T of $\mathbb{R}^n$ into $\mathbb{R}^n$ satisfying conditions (a) and (b) corresponds to a certain matrix A. For the sake of brevity we denote the matrix and the corresponding transformation by the same letter, A. Given an arbitrary norm $\|\cdot\|$ on $\mathbb{R}^n$ (or $\mathbb{C}^n$), we can define the norm $\|\cdot\|$ of an arbitrary matrix (or linear transformation) A by

$$\|A\| = \max_{\|x\|=1} \|Ax\|.$$

This norm has the following properties:

(a) $\|A + B\| \le \|A\| + \|B\|$ for any matrices A, B.

(b) $\|Ax\| \le \|A\|\|x\|$ for any matrix A and any vector x.

(c) $\|AB\| \le \|A\|\|B\|$ for any matrices A, B.

1.1.4 For an arbitrary matrix A we say that a nonzero vector x is an eigenvector of A if $Ax = kx$ for some number k. In this case k is called an eigenvalue of the matrix A.

If $Ax = kx$, then $(A - kI)x = 0$ and $A - kI$ is a degenerate matrix; therefore, its determinant vanishes, that is, $\det(A - kI) = 0$. The function $P(k) = \det(A - kI)$ is called the characteristic polynomial of the matrix A. It is a polynomial of degree n, where n is the dimension of the vector space $\mathbb{R}^n$.

For example, for a 2 × 2 matrix

$$A = \begin{pmatrix} a & b \\ c & d \end{pmatrix}$$

we have

$$P(k) = \det(A - kI) = \det\begin{pmatrix} a - k & b \\ c & d - k \end{pmatrix}$$
$$= (a - k)(d - k) - bc = k^2 - (a + d)k + (ad - bc).$$

This is a polynomial of degree 2. The eigenvalues of A are precisely the roots of its characteristic polynomial; therefore, there are no more than n eigenvalues. Even for a real matrix A, its characteristic polynomial may have complex roots. Thus an eigenvalue k of a real matrix A may be a complex number and the corresponding eigenvector x may have complex coordinates.

1.1.5 To construct converging iterative processes and to study their rates of convergence, the aforementioned notions, such as the norm of a matrix or the set of its eigenvalues, can be very useful. For example, if $\|A\| = \alpha < 1$, then $\lim_{\sigma\to\infty}\|A^\sigma x\| = 0$ for any vector x.

Moreover, the rate of convergence can be estimated by means of α, namely, $\|A^\sigma x\| \le \alpha^n\|x\|$. Generally speaking, for any eigenvalue k of a matrix A, we have $|k| \le \|A\|$. The converse is not true.

For example, if

$$A = \begin{pmatrix} 0 & 5 \\ 0 & 0 \end{pmatrix}$$

then $\|A\| = 5$ but has only one eigenvalue equal to 0.

Nevertheless, if for any eigenvalue k of a matrix A we have $|k| < \alpha$ where $\alpha < 1$, then $\lim_{\sigma\to+\infty}\|A^\sigma x\| = 0$. Moreover, $\|A^\sigma x\| \le C\alpha^\sigma\|x\|$ for some constant C. So the rate of convergence of $A^\sigma x$ to zero is determined by the maximal eigenvalue of A rather than by the norm of A. On the other hand, it is often far easier to estimate the norm of a matrix than to estimate its maximal eigenvalue k_{max} (except for the trivial observation that $|k_{max}| \le \|A\|$). For example, if $A = (a_{ij})$ has the property $\max_j \Sigma_i |a_{ij}| = \alpha < 1$, then $\|A\| \le \alpha$ in the

ℓ_1 norm. Indeed, if $\|x\| = 1$, then $\Sigma_i |x_i| = 1$. Therefore, $\|Ax\| = \Sigma_i |\Sigma_j a_{ij}x_j| \le \Sigma_i \Sigma_j |a_{ij}||x_j| = \Sigma_j (|x_j| \Sigma_i |a_{ij}|) \le \Sigma_j \alpha|x_j| = \alpha$. So $\|A^\sigma x\| \le \alpha^\sigma \|x\|$. The maximal eigenvalue k_{max} of A may provide better estimates for the rate of convergence, but it would require far more calculations to approximate this eigenvalue.

1.1.6 The maximal eigenvalue of a matrix A and its associated eigenvector can be obtained with the help of a simple iterative procedure. Assume that the maximal eigenvalue k of A is a simple root of the characteristic polynomial of A, and that all other eigenvalues of A are less than $|k|$ in absolute value. Note that in this case k must be real (otherwise, the conjugate of k will be another eigenvalue of A with the same absolute value). These conditions are satisfied, for example, if A is a matrix with positive entries.

Start with an arbitrary vector x^0 and define a sequence of vectors x by the formulas

$$x^{\sigma+1} = \frac{Ax^\sigma}{\|Ax^\sigma\|}.$$

Obviously,

$$x^{\sigma+1} = \frac{A^{\sigma+1}x^0}{\|A^{\sigma+1}x^0\|},$$

and for $\sigma > 0$ we have $\|x^\sigma\| = 1$.

Assume for simplicity that all the eigenvalues of A are simple roots of the characteristic polynomial of A. Let $k_1 = k, k_2, k_3, \ldots, k_n$ be the eigenvalues of A and $|k| > |k_2| \ge |k_3| \ge \cdots \ge |k_n|$. Let $p_1, \ldots, p_n$ be associated eigenvectors. Let $x^0 = a_1p_1 + a_2p_2 + \cdots + a_np_n$.

Since x^0 is arbitrary, we can assume that $a_1 \ne 0$. Replacing, if necessary, p_1 by $\pm p_1/\|p_1\|$, we can assume that $a_1 > 0$ and $\|p_1\| = 1$. Then

$$A^\sigma x^0 = \sum_i a_iA^\sigma p_i = \sum_i a_ik_i^\sigma p_i = k^\sigma\left[a_1p_1 + \sum_{i=2}^n a_i\left(\frac{k_i}{k}\right)^\sigma p_i\right].$$

Since $|k_i/k| < 1$ for $i \geq 2$, we have

$$\lim_{\sigma\to\infty} \frac{A^\sigma x^0}{k^\sigma} = a_1 p_1$$

and

$$\lim_{\sigma\to\infty} x^\sigma = \lim_{\sigma\to\infty} \frac{A^\sigma x^0}{\|A^\sigma x^0\|} = \lim_{\sigma\to\infty} \frac{A^\sigma x^0/k^\sigma}{\|A^\sigma x^0/k^\sigma\|} = \frac{a_1 p_1}{\|a_1 p_1\|} = p_1.$$

Approximations of k are given by the formulas $\alpha^\sigma = (Ax^\sigma, x^\sigma)$. This is why

$$\lim_{\sigma\to\infty} \alpha^\sigma = \lim_{\sigma\to\infty} (Ax^\sigma, x^\sigma) = (Ap_1, p_1) = k\|p_1\|^2 = k.$$

1.2 METHODS OF UNCONSTRAINED OPTIMIZATION

In this section we review methods for solving the problem

$$\max F(x), \qquad x \in \mathbb{R}^n, \tag{1.2.1}$$

where $\mathbb{R}^n$ denotes the n-dimensional linear space, which is the set of all n-dimensional vectors $x = (x_1, \ldots, x_n)$. In what follows it is assumed that the function F(x) is differentiable; for Newton and quasi-Newton methods it is further assumed that F(x) has continuous second derivatives.

Some problems, which are not unconstrained optimization problems formally, can be recast in the form of equation (1.2.1). For example, finding the solution (if one exists) to the system of equations

$$f_g(x) = 0, \qquad g = 1, \ldots, m, \qquad x \in \mathbb{R}^n \tag{1.2.2}$$

is often equivalent to solving the problem (1.2.1):

$$\max F(x), \qquad F(x) = -\sum_{g=1}^{m} f_g^2(x), \qquad x \in \mathbb{R}^n. \tag{1.2.3}$$

For example, problems (1.2.2) and (1.2.3) are equivalent if for any vector x that is a local maximum of F(x), the rank of the gradient matrix $[\partial f_g(x)/\partial x_q]$ is equal to m. Indeed, if x* is a local maximum of F(x), then x* satisfies

$$\frac{\partial F(x^*)}{\partial x_q} = 2 \sum_g f_g(x^*) \frac{\partial f_g(x^*)}{\partial x_q} = 0, \qquad q = 1, \ldots, n.$$

If the gradient matrix is of full rank, then x* satisfies (1.2.2). In general, however, this need not be true, so the local maximum should always be checked to see if it satisfies (1.2.2). Note that if all the $f_g(x)$ are differentiable, then F(x) is also differentiable.

In particular, the problem of solving an arbitrary linear system of equations

$$Ax = b, \tag{1.2.4}$$

where A is a matrix and b and x are vectors of comparable dimensions, is equivalent to the problem of solving

$$-\|Ax - b\|^2 \to \max, \qquad x \in \mathbf{R}^n, \tag{1.2.5}$$

provided that (1.2.4) has a solution. Therefore,

$$\|Ax - b\|^2 = (Ax, Ax) - 2(Ax, b) + (b, b). \tag{1.2.6}$$

Indeed, the maximum of the quantity $-\|Ax - b\|^2$ is equal to zero, and the norm of the vector (and hence the square of the norm) is equal to zero only when the vector itself is equal to zero.

At first glance the problem (1.2.3) may appear to be more complicated than the problem (1.2.2). However, reformulating (1.2.2) into (1.2.3) and solving this equivalent problem is the simplest and most widely used method of solution for (1.2.2). Conversely, in many particular cases [e.g., when F(x) is convex and m = n], solving (1.2.1) reduces to the problem of solving the following system of equations:

$$\frac{\partial F(x)}{\partial x_i} = 0, \qquad i = 1, \ldots, n, \qquad x \in \mathbf{R}^n. \tag{1.2.7}$$

Virtually all of the methods for solving (1.2.1) are iterative methods. Moreover, most of the solution methods construct a sequence of approximations $x^0, x^1, \ldots$, such that the sequence $F(x^\sigma)$ increases monotonically [i.e., $F(x^\sigma) > F(x^{\sigma-1})$ unless x^σ obtains the maximum of $F(x)$]. Such iterative methods are called "relaxation methods."

The most common iterative relaxation algorithms proceed as follows. Let $x^{\sigma-1}$ be the known approximate solution obtained after iteration $\sigma - 1$. The vector of directional increments (VDI) $\Delta x^{\sigma-1}$ is also given (the coordinates of $\Delta x^{\sigma-1}$ are allowed to be negative). Then an updated approximation x^σ is chosen based on given criteria from all vectors of the form $x^{\sigma-1} + \alpha^\sigma \Delta x^{\sigma-1}$, where α^σ is an arbitrary real number. Usually, $\Delta x^{\sigma-1}$ has a unit norm (i.e., $\|\Delta x^{\sigma-1}\| = 1$). The various methods for solving (1.2.1) actually differ only in their choice of Δx^σ and α^σ.

Algorithms in which α^σ is chosen to maximize the value of the function F from among all vectors of the form $x^{\sigma-1} + \alpha^\sigma \Delta x^{\sigma-1}$ will be referred to as algorithms with exact one-dimensional maximization. In these algorithms at each iteration we solve the problem

$$\bar{F}(\alpha) = F(x^{\sigma-1} + \alpha \Delta x^{\sigma-1}) \to \max; \qquad \alpha \in \mathbb{R}^1. \tag{*}$$

For example, when maximizing a quadratic function $Q(x) = (Ax,x) + (b,x)$, for given values of $x^{\sigma-1}$ and $\Delta x^{\sigma-1}$ at iteration $\sigma - 1$, we can find the value of $\alpha\Delta x^{\sigma-1}$ that maximizes the function F. By substituting Q(x) for F(x) in (*), the optimal step size solves the problem:

$$[(Ax^{\sigma-1},x^{\sigma-1}) + \alpha(Ax^{\sigma-1},\Delta x^{\sigma-1}) + \alpha(A\Delta x^{\sigma-1},x^{\sigma-1})$$
$$+ \alpha^2(A\Delta x^{\sigma-1},\Delta x^{\sigma-1})] + (b,x^{\sigma-1}) + \alpha(b,\Delta x^{\sigma-1}) \to \max.$$

By making the substitution

$$a_1 = (Ax^{\sigma-1},\Delta x^{\sigma-1}) + (A\Delta x^{\sigma-1},x^{\sigma-1}) + (b,\Delta x^{\sigma-1});$$
$$a_2 = (A\Delta x^{\sigma-1},\Delta x^{\sigma-1});$$
$$a_3 = (Ax^{\sigma-1},x^{\sigma-1}) + (b,x^{\sigma-1}),$$

the problem above can be rewritten as

$$\alpha^2 a_2 + \alpha a_1 + a_3 \to \max.$$

If a solution exists (it exists when $a_2 < 0$), it is given by

$$\alpha = -\frac{a_1}{2a_2}.$$

The end result of fixing the ratio of the increments in the original problem is that at each iteration of the algorithm we solve only a one-dimensional extremum problem instead of the multidimensional problem in the original formulation.

Given such an approach for solving (1.2.1), the choice of the VDI assumes great importance. The nature of this choice is as follows. Assume that an initial approximation x^0 is given together with information about the derivatives of the objective function at this point. From this information we seek to find a VDI that maximizes the value of the objective function. Generally, there exists a VDI that yields the optimum in one step. However, the information concerning the behavior of the objective function in a neighborhood of x^0 is usually sufficient to give only a rough estimate of the optimal value for large increments. Since it is not possible to obtain an optimal solution in one step, we use iterative procedures to improve the estimates.

Suppose that we can determine the value of the function $F(x^0 + \Delta x^0)$ in a neighborhood of x^0 only for values of the vector increments Δx^0 with norms satisfying $\|\Delta x^0\| \leqq \delta$ for some small positive number δ. We select as the unknown direction of increase the vector Δx^0 that gives the maximum increase in the value of the function. This is precisely what is done in gradient methods.

The gradient vector $\nabla F(x^0)$ (i.e., the vector of partial derivatives) is the vector where $x^0 + \alpha\nabla F(x)$ has the maximal increase for sufficiently small α in a neighborhood of x^0. However, the direction that yields the maximum increase in a neighborhood of x^0 may yield poor results when well outside this neighborhood.

The mathematical basis of gradient methods becomes clearer if we consider the Taylor series approximation of a smooth function $F(x)$

in a neighborhood of a point x^0:

$$F(x) = F(x^0) + (\nabla F(x^0), (x - x^0)) + \frac{1}{2} (\nabla_2 F(x^0)(x - x^0), (x - x^0)) + \cdots, \qquad (1.2.8)$$

where

$$\nabla F(x^0) = \left\{\frac{\partial F}{\partial x_1} (x^0), \ldots, \frac{\partial F}{\partial x_n} (x^0)\right\}$$

is the gradient vector of F(x) at x^0 and

$$\nabla_2 F(x) = \left\{\frac{\partial^2 F(x^0)}{\partial x_i \, \partial x_j} (x^0)\right\}$$

is the matrix of second derivatives of F(x) at x^0. It is clear from (1.2.8) that gradient methods choose the direction of increase using the very coarse approximation $F(x) = \Phi(x) = F(x^0) + (\nabla F(x^0), (x - x^0))$, and that a direction is chosen from x^0 that will yield the most rapid increase in $\Phi(x)$. That direction coincides with the vector $\nabla F(x^0)$.

In choosing a VDI it would seem desirable to use a less coarse approximation to F(x) by using further terms in the expansion (1.2.8). If the function to be maximized, F(x), is a quadratic function of the form $F(x) = (Ax,x) + (b,x) + c$, then when the maximum exists, it can be found by solving a system of linear equations,[1] $(A + A')x = -b$, since at this point and only at this point does the gradient of F(x) equal 0.

Maximization methods that use second derivatives use this property of the gradient of a quadratic function. In these methods a quadratic function is constructed based on the local behavior of F(x) such that in a small neighborhood of x^0 this function provides the best quadratic approximation to F(x). From (1.2.8) it follows that the best quadratic approximation is given by

---

[1] The matrix A' here and later denotes the transpose of matrix A.

$$\psi(x) = F(x^0) + (\nabla F(x^0), (x - x^0)) + \frac{1}{2} (\nabla_2 F(x^0)(x - x^0), (x - x^0)). \qquad (1.2.9)$$

We then select a vector $\widetilde{\Delta x}^0$ such that the function $\psi(x)$ is maximized at $x^0 + \widetilde{\Delta x}^0$. Since $\nabla_2 F(x^0)$ is a symmetric matrix [i.e., $(\nabla_2 F(x^0))' = \nabla_2 F(x^0)$], it follows that $\widetilde{\Delta x}^0$ is determined by solving the system of equations

$$\nabla_2 F(x^0)\widetilde{\Delta x}^0 = -\nabla F(x^0).$$

If the inverse Hessian matrix $(\nabla_2 F(x^0))^{-1}$ is known, $\widetilde{\Delta x}^0$ can be obtained by matrix multiplication as

$$\widetilde{\Delta x}^0 = (\nabla_2 F(x^0))^{-1}(-\nabla F(x^0)).$$

When $\widetilde{\Delta x}^0$ has been determined, the VDI is usually taken to be $\Delta x^0 = \widetilde{\Delta x}^0/\|\widetilde{\Delta x}^0\|$. A detailed discussion of these methods can be found in Gill and Murray (1974).

Unfortunately, in practice these methods are usually not capable of solving large-dimensional problems since at each iteration it would require solving a high-order linear system of equations. Also, there can be computational difficulties in calculating $\nabla_2 F(x^0)$. Because of these problems with gradient methods, at the present time the so-called quasi-Newton or variable metric algorithms are in wide use. In these algorithms the Hessian matrix $\nabla_2 F(x^0)$ or its inverse is adjusted in the course of the computations. This adjustment only uses the gradient evaluated at points from previous iterations. Thus the quasi-Newton algorithms require the storage of $n \times n$ matrices, where n is the number of variables in the problem; and for iterations where σ is large, the matrices to be stored may not be sparse. Therefore, quasi-Newton algorithms can also not readily be used to solve large-dimensional problems. For such problems it is necessary to use various methods for compact storage of matrices during computation.

In the algorithms discussed above, the VDIs are selected based on the information available about the local behavior of the objective function in a neighborhood of the sequence of points generated by the algorithm. However, it is also possible to select in advance a fixed set of vectors and then select the VDIs from this set in a specified manner. This procedure, for example, is used in the different variants of coordinate descent methods, where the coordinate axes are alternatively being used as the VDI. As a rule, the latter algorithms are considerably less effective than the algorithms that choose the directions according to the information provided by the sequence of points generated by the algorithm. The one exception to this rule, and it is an exception that is important both practically and theoretically, is the conjugate gradient method.

Before describing the conjugate gradient algorithm in more detail, we will review some necessary facts concerning systems of conjugate vectors. The vectors $v_1, \ldots, v_k$ are said to be conjugate with respect to the matrix A if $(Av_i, v_j) = 0$ for all $i \neq j$ and $v_i \neq 0$ for all i. Let H denote an $n \times n$ symmetric matrix (i.e., $H' = H$). Then there exist n nonzero linearly independent vectors that are conjugate with respect to H.

Moreover, any set of k conjugate vectors can be expanded to a system of n conjugate vectors. Also, any set of conjugate vectors is linearly independent. Vectors conjugate with respect to H have the following remarkable property. Let $v_1, \ldots, v_k$ be k vectors conjugate with respect to H. Consider a quadratic form $Q(x) = -(1/2)(Hx,x) + (b,x)$, where b is an arbitrary n-dimensional vector. Construct a sequence of vectors $x^0, x^1, \ldots, x^k$ as follows. Choose an arbitrary vector x^0. Consider the line passing through x^0 parallel to v_1 and find a point x^1 on this line where Q(x) attains its maximum value. Next find a point x^2 on the line passing through x^1 and parallel to v_2 at which Q(x) reaches its maximum value on this line. Continuing this procedure, we move from x^i along a line parallel to v_{i+1} until we reach a point x^{i+1}, where Q(x) attains its maximum value on this line. Finally, we obtain a point x^k. The remarkable property of the vectors $v_1, \ldots, v_k$ is that $Q(x^k)$ equals the maximum value of Q(x) on the k-dimensional space spanned by $v_1, \ldots, v_n$ and passing through x^0. For example, if v_1, v_2 are conjugate with respect to H and x^0 is an arbitrary point, we find x^1 as a point of maximum of Q(x) on the line through x^0 parallel to v_1. Next, we find x^2 as a point of maximum of Q(x) on the line through x^1 parallel to v_2. Then the resulting point x^2 is a point of maximum of Q(x) on the plane passing through x^0 and parallel to v_1 and v_2.

From a computational point of view the main advantage of conjugate vectors compared with nonconjugate ones is that the problem of maximizing a quadratic form is reduced to a sequence of one-dimensional maximization problems. An example of conjugate vectors for the case is given in Figure 1.1.

While constructing algorithms based on the notion of conjugacy, we have to take into account that in the beginning of computations we do not have conjugate vectors. Therefore, in addition to one-dimensional maximizations, such algorithms should enable us to find conjugate vectors.

One of the most widely used conjugate directions algorithms is the conjugate gradient algorithm. In this algorithm the conjugate directions are constructed successively. The kth conjugate vector v_k is determined from the formula

$$v_k = \nabla F(x^k) + \frac{(\nabla F(x^k), \nabla F(x^k))}{(\nabla F(x^{k-1}), \nabla F(x^{k-1}))} v_{k-1}, \quad k = 1, \ldots, n, \qquad (1.2.10)$$

with $v_0 = 0$. For the quadratic function $Q(x) = -(1/2)(Hx,x) + (b,x)$ with H a symmetric positive-definite matrix, the directions $(v_1, \ldots, v_n)$ generated by the conjugate gradient algorithm are conjugate with respect to H. Therefore, the conjugate gradient algorithm will maximize a quadratic function in no more than n steps, and approximations sufficient for practical purposes are usually reached after a relatively few iterations.

For certain special classes of unconstrained optimization problems, algorithms that are not relaxation algorithms are being used to find solutions. Thus to solve the problem

$$F_i(x) = 0, \qquad i = 1, \ldots, n, \qquad (1.2.11)$$

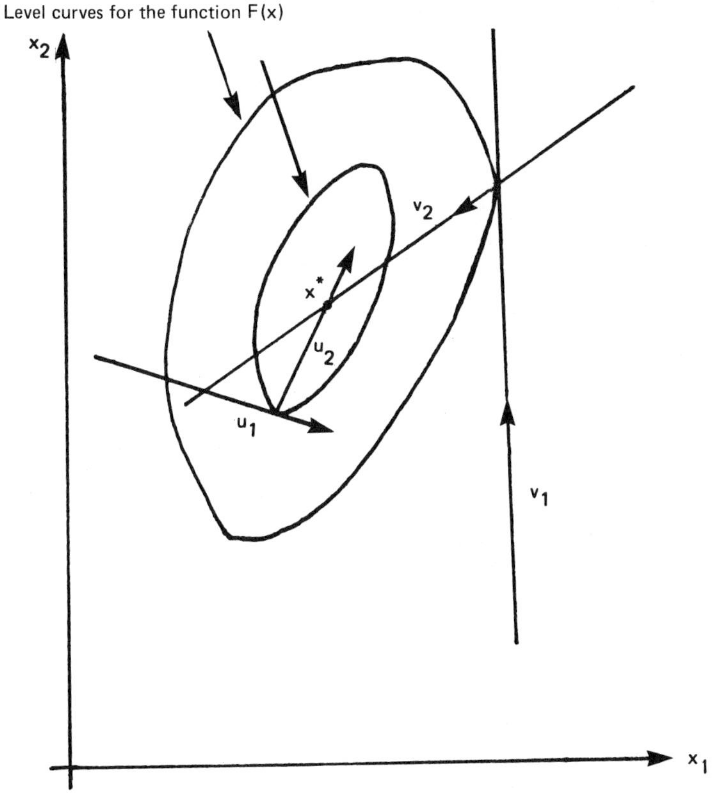

Figure 1.1. Two families of conjugate vectors for the case of quadratic function $Q(x)$, $n = 2$; the vector v_1 is conjugate to v_2, and the vector u_1 is conjugate to u_2 (x^* is a maximum point).

which, as noted above, can be reformulated as an unconstrained optimization problem, we frequently use contraction mapping algorithms. In such algorithms, x^σ is found by solving the system of equations

$$g_i(x^\sigma) = f_i(x^{\sigma-1}), \qquad i = 1, \ldots, n, \tag{1.2.12}$$

where

$$g_i(x) - f_i(x) = F_i(x), \qquad i = 1, \ldots, n. \tag{1.2.13}$$

When equation (1.2.13) is satisfied, $g_i(x^*) = f_i(x^*)$ for the solution of the original system x^*. Also, if $x^\sigma = x^{\sigma-1}$, then

$F(x^\sigma) = g_i(x^\sigma) - f_i(x^\sigma) = 0$, and therefore $x^\sigma = x^*$. These algorithms will converge if the function $g^{-1}(f(x))$ is a contraction mapping, where $g^{-1}(x) = (g_1^{-1}(x), \ldots, g_n^{-1}(x))$, $f(x) = (f_1(x), \ldots, f_n(x))$, and $g^{-1}(x)$ is the inverse function of $g(x)$. Recall that a mapping $x \to G(x)$ is said to be contracting if for any x_1 and x_2 the inequality $\|x_1 - x_2\| > \|G(x_1) - G(x_2)\|$ is satisfied.

Such widely known algorithms for solving linear systems as successive approximations, simple iteration, the Gauss-Seidel method, and the Nekrasov method are all contraction mapping algorithms. In each of these algorithms the general problem can be expressed in the form

$$x = Ax + b, \qquad A = (a_{ij}), \qquad b = (b_i), \qquad i, j = 1, \ldots, n. \tag{1.2.14}$$

In the successive approximations algorithm x^σ is found from the formula

$$x^\sigma = Ax^{\sigma-1} + b.$$

The Gauss-Seidel algorithm successively solves for the coordinates of the vector x^σ. Assume that for iteration σ the coordinates x_k^σ of x^σ have been determined for $k < i$. Then

$$x_i^\sigma = \sum_{j=1}^{i-1} a_{ij}x_j^\sigma + \sum_{j=i}^{n} a_{ij}x_j^{\sigma-1} + b_i.$$

Simple (Jacobi) iterations solve for x_i^σ from the formula

$$x_i^\sigma = \frac{1}{1 - a_{ii}} \left(\sum_{j \neq i} a_{ij}x_j^{\sigma-1} + b_i \right).$$

The Nekrasov algorithm combines simple and Gauss-Seidel iterations. As in the Gauss-Seidel algorithm, assume that at iteration σ the coordinates x_k have been determined for all $k < i$. Then x_i^σ is determined from the formula

$$x_i^\sigma = \frac{1}{1 - a_{ii}} \left(\sum_{j=1}^{i-1} a_{ij}x_j^\sigma + \sum_{j=i+1}^{n} a_{ij}x_j^{\sigma-1} + b_i \right).$$

There are also block analogs of simple iterations and of Gauss-Seidel and Nekrasov algorithms. Assume that the variables are partitioned into L disjoint sets. Also, let G_t $(t = 1, \ldots, L)$ denote the set of indices belonging to the tth partition. The block analog of simple iterations algorithm works as follows.

The component values of x_i^σ for $i \in G_t$ are found by solving the system of equations

$$x_i^\sigma = \sum_{j \in G_t} a_{ij} x_j^\sigma + \sum_{j \notin G_t} a_{ij} x_j^{\sigma-1} + b_i.$$

Block versions of the Gauss-Seidel and Nekrasov algorithms are constructed analogously.

In each of the algorithms discussed above, the x^σ are obtained from a linear transformation of $x^{\sigma-1}$ of the form

$$x^\sigma = Bx^{\sigma-1} + c,$$

where B is a matrix and c is a vector, which are determined from the values of A and b in the related procedure.

Let us discuss the convergence of the iterative process $x^\sigma = Bx^{\sigma-1} + c$ to the solution x^* of the equation $x = Bx + c$. Let $v^\sigma = x^\sigma - x^*$; then $x^\sigma = x^* + v^\sigma$ and $x^* + v^\sigma = Bx^* + Bv^{\sigma-1} + c$. Since $x^* = Bx^* + c$, we have $v^\sigma = Bv^{\sigma-1}$. It is well known that v^σ converges to zero (and therefore x^σ converges to x^*) if $\|B\| < 1$, where $\|\cdot\|$ is an arbitrary norm. The condition $\|B\| < 1$ can be replaced by a weaker condition. Recall that a number λ is said to be an eigenvalue of the matrix B if for some nonzero vector w we have $Bw = \lambda w$. In this case w is called an eigenvector corresponding to the eigenvalue λ. A necessary and sufficient condition for the process $x^\sigma = Bx^{\sigma-1} + c$ to converge for any initial vector x^0 is that $|\lambda| < 1$ for every eigenvalue λ of the matrix B. This condition is satisfied, for example, if $\|B\| < 1$. Moreover, the rate of convergence is determined by the eigenvalue λ with maximal absolute value.

1.3 METHODS OF CONSTRAINED OPTIMIZATION

By the phrase "methods of constrained optimization" we mean methods for solving mathematical programming problems of the form

$$f(x) \to \max; \qquad x \in Q, \tag{1.3.1}$$

where Q is a subset of $\mathbb{R}^n$. The problem given by (1.3.1) differs from (1.2.1) in that the search for the maximum of f(x) is carried out subject to the additional constraint that $x \in Q$—hence the name "constrained optimization."

Often, the set Q is defined as the intersection of a finite number of other sets: $Q = \cap_{j=1}^{m} Q_j$. This means that x must belong to each of the constraint sets Q_j, j = 1, ..., m. The sets Q_j are usually defined by systems of equalities and inequalities of the form $Q_j = \{x | g_i^j(x) \leq 0,\ i = 1, \ldots, m_1;\ g_i^j(x) = 0,\ i = m_1 + 1, \ldots, m\}$. For example, if $Q_j = \{x | g_i^j(x) \leq 0,\ i = 1, \ldots, n\}$, when $g_i^j(x) = -x$, then $Q_j \subset \mathbb{R}_+^n$, where $\mathbb{R}_+^n$ is the subset of $\mathbb{R}^n$ comprised of all non-negative n-dimensional vectors. When the constraint is written as $x \in Q$ with Q unspecified, this denotes that the structure of Q is not important to the point under discussion.

At the present time, the most widely used methods for solving (1.3.1) are the simplex algorithm and its modifications. However, since we do not use these algorithms for constructing iterative aggregation algorithms, they will not be discussed further.

Another important class of algorithms for solving mathematical programming problems involve reducing the original problem to a sequence of optimization problems that have a simpler structure than that of the original problem. Penalty function algorithms and barrier function algorithms are examples of this type of algorithm.

These algorithms reduce the solution of the problem

$$\begin{aligned} &f(x) \to \max; \\ &g_i(x) \leq 0, \qquad i = 1, \ldots, m_1; \\ &g_i(x) = 0, \qquad i = m_1 + 1, \ldots, m, \end{aligned} \tag{1.3.2}$$

to that of solving the sequence of problems

$$f(x) - \Sigma_i \, u_i(q_i^\sigma, g_i(x)) \to \max; \qquad x \in Q, \tag{1.3.3}$$

where the q_i^σ are parameters of the process that are increasing in σ and $u_i(q,t)$ are functions of the variables q and t.

Penalty function algorithms generally require that the functions $u_i(q,t)$ satisfy the following conditions:

(a) $u_i(q,t) = 0$ if $t \leqq 0$, $i \leqq m_1$ for all q.
(b) $u_i(q,t) = 0$ if $t = 0$, $i > m_1$ for all q.
(c) $\lim_{q\to\infty} u_i(q,t) = \infty$ if $t > 0$, $i \leqq m_1$.
(d) $\lim_{q\to\infty} u_i(q,t) = \infty$ if $t \neq 0$, $i > m_1$.

Usually, the u_i are also assumed to be monotone in the first argument and strictly positive:

(e) $u_i(q_1,t) \geqq u_i(q_2,t)$ if $q_1 > q_2$.
(f) $u_i(q,t) > 0$ if $q > 0$, $t > 0$, and $i \leqq m_1$,
(g) $u_i(q,t) > 0$ if $q > 0$, $t \neq 0$, and $i > m_1$.

In penalty function algorithms the most widely used forms for the u_i are[1]

$$\begin{aligned} u_i(q,t) &= q(t_{(+)})^2, \qquad i = 1, \ldots, m_1, \\ u_i(q,t) &= qt^2, \qquad i = m_1 + 1, \ldots, m. \end{aligned} \tag{1.3.4}$$

Barrier functions are used to solve optimization problems with only inequality constraints [i.e., $m_1 = m$ in (1.3.1)]. In these algorithms the functions $u_i(q_i, g_i(x))$ are chosen to set a threshold at the boundary of the feasible region of the problem; in other words, for a smooth barrier function they prevent us from violating the constraints (1.3.2) when solving (1.3.3) by one of the methods outlined in Section 2.1. Also, the u_i must be chosen so that their influence within the feasible region decreases as q_i tends to infinity. In other words, for all $b < 0$ it must be true that

---

[1]Here and later, $a_{(+)} = \begin{cases} 0 & \text{if } a \leq 0; \\ a & \text{if } a > 0. \end{cases}$

$\lim_{q_i \to \infty} u_i(q_i, b) = 0$. An example of a function satisfying these conditions is $u(q,t) = (1/q) \ln(-t)$.

Penalty function methods solve (1.3.2) as follows. Given initial values for the parameters $q_i^0 > 0$, $i = 1, \ldots, m$, we first solve (1.3.3) with $q_i = q_i^0$. Denote this solution by x^1. The functions g_i, $i = 1, \ldots, m$, are then evaluated at x^1, and we compute $\Delta_i^1 = g_i(x^1)$ for $i = 1, \ldots, m_1$ and $\Delta_i^1 = g_i(x^1)_{(+)}$ for $i = m_1 + 1, \ldots, m$. Based on criteria determined before starting the algorithm, if any of the Δ_i are too large, the appropriate q_i is increased in value. Usually, the numbers α and β are chosen to satisfy $1 > \alpha > 0$ and $\beta > 1$, and q_i^σ is chosen to satisfy $q_i^\sigma = q_i^{\sigma-1}$ if $\Delta_i^\sigma \le \alpha\Delta_i^{\sigma-1}$ and $q_i^\sigma = \beta q_i^{\sigma-1}$ if $\Delta_i^\sigma > \alpha\Delta_i^{\sigma-1}$. The sequence of approximate solutions x^0, x^1, ... generated by the penalty function method will almost always converge to the solution of the problem (1.3.2).

When solving (1.3.2) using barrier functions, the numbers β and q_i^0 are usually chosen to satisfy $\beta > 0$ and $q_i^0 > 0$, and for $k = 0, 1, 2, \ldots$ we solve (1.3.3) with $q_i = \beta q_i^0$, $i = 1, \ldots, m$. The solution of (1.3.2) is taken to be the limit of the sequence $x^0, x^1, \ldots$.

The functions used as barrier functions are assumed to be nearly discontinuous. This fact greatly limits their applicability. Fiacco and McCormick (1968) provide a detailed description of both penalty function and barrier function methods for solving (1.3.2).

The third group of methods for solving (1.3.2) consists of methods that transform the original problem into that of finding the saddle point of some function over a set of relatively simple structure. These algorithms include the methods (considered in Part III) that use classical or augmented Lagrangian functions, we well as the solution techniques that transform the original problem into an equivalent game. The latter methods are described in detail in Belen'ky and Volkonsky (1974).

Recall that the Lagrangian function for the problem

$$
\begin{aligned}
&f(x) \to \max; \\
&g_i(x) \le 0, \qquad i = 1, \ldots, m_1; \\
&g_i(x) = 0, \qquad i = m_1 + 1, \ldots, m,
\end{aligned}
\tag{1.3.5}
$$

is defined as

$$F(x,y) = f(x) - \Sigma_i\, y_i g_i(x), \tag{1.3.6}$$

where $y = (y_i)$. The variables y_i are called the Lagrangian multipliers or dual variables of the problem. If the function f(x) in (1.3.5) is convex and the functions $g_i(x)$ are concave, there exists a vector $y^* = (y_i^*,\ i = 1, \ldots, m)$ such that (x^*,y^*) is the saddle point of F(x,y) in the set $\{y_i \geqq 0,\ i = 1, \ldots, m_1\}$, where x^* is defined as the solution to (1.3.5). It is this property of the Lagrangian function that makes it useful in solving problems of the form (1.3.5).

Unfortunately, the Lagrangian function often does not possess the properties necessary to construct effective algorithms for solving (1.3.5) (see Golstein, 1972). In augmented Lagrangian methods, the function F(x,y) is modified in order to have these desirable properties. Usually, in the algorithms in this book the Lagrangian function is modified by the introduction of an additional penalty term of the form

$$M(x,y) = f(x) - \sum_{i=1}^{m_1} \frac{1}{2q_i}[(y_i + g_i(x))_{(+)}]^2 + \sum_{i=1}^{m_1} \frac{y_i^2}{2q_i} - \sum_{i=m_1+1}^{m} y_i g_i(x) - \sum_{i=m_1+1}^{m} \frac{q_i}{2} g_i^2(x), \tag{1.3.7}$$

where q_i is a parameter of the process.[1]

---

[1]For the inequalities the modified Lagrangian function is

$$f(x) - \sum_{i=1}^{m_1} ((y_i + g_i(x))_{(+)})^2 + \sum_{i=1}^{m_1} \frac{y_i^2}{2q_i}.$$

For the equalities the modified Lagrangian function is

$$f(x) - \sum_{i=m_1+1}^{m} y_i g_i(x) - \sum_{i=m_1+1}^{m} \frac{q_i}{2} g_i^2(x).$$

In (1.3.7) both these functions are used.

The most popular algorithms that use (1.3.7) for solving (1.3.2) use successive adjustment of the y variables. For a fixed value of $y^{\sigma-1}$ we solve the problem

$$M(x,y^{\sigma-1}) \to \max; \qquad x \in Q, \tag{1.3.8}$$

whose solution is denoted by x^{σ}. Then y^{σ} is found from

$$\begin{aligned} y_i^{\sigma} &= (y_i^{\sigma-1} + q_i g_i(x^{\sigma}))_{(+)}, \qquad i = 1, \ldots, m_1; \\ y_i^{\sigma} &= y_i^{\sigma-1} + q_i g_i(x^{\sigma}), \qquad i = m_1 + 1, \ldots, m. \end{aligned} \tag{1.3.9}$$

By constructing the algorithm in this manner it is no longer necessary to choose the q to tend to infinity in order for the algorithm to converge to the solution [the equality-constrained case is discussed in Polyak and Tretyakov (1972); the general case is discussed in Tretyakov (1973) and Fletcher (1973)]. This improves the convergence of the algorithm.

It is our opinion that penalty function algorithms and algorithms that use Lagrangian functions modified by a penalty function best reflect the economic nature of the problems under study in this book. In real-life problems the constraints are rarely rigidly binding. Replacing the constraint with a quadratic penalty term with the proper choice of the penalty nearly always yields a more realistic formulation of the real problem than does a formulation that requires that the constraints be strictly observed. Although it is true that many references to the problems encountered when using penalty function algorithms—namely, that the penalty coefficients must tend to infinity and it is very time consuming to solve the problem when there are large penalties—in economic problems these difficulties are often of little practical importance since it is not necessary to obtain the solution with such accuracy.

Moreover, in penalty function or modified Lagrangian algorithms it is possible to put only those constraints that are flexible into the penalty term while leaving the other constraints as part of the problem. Modified Lagrangian algorithms usually generate quadratic programming problems that are only slightly more time consuming to solve than a linear programming problem with the same dimensions. Moreover, finding the minimum of a function over a set defined by a small number of linear constraints requires approximately as many operations as solving the unconstrained minimization of the same function.

Finally, we note that the so-called "projection methods" have been shown to be effective at solving maximization problems with linear constraints. The idea of these algorithms is to apply to the constrained problems certain methods that were discussed above for the unconstrained problems; namely, relaxation methods that use one-dimensional optimization. As in the former case, projection algorithms construct a VDI and then minimize along this direction. However, now the VDI must be chosen with the additional constraint that it does not go beyond the set of feasible points. Usually, this is accomplished by projecting a specially selected VDI onto the set of feasible points and then minimizing along that projection. Recall that for a given vector $u = (u_1, \ldots, u_n)$, its projection onto a set of equations

$$\sum_{j=1}^{n} a_{ij} x_j = b_i, \qquad i = 1, \ldots, m, \tag{1.3.10}$$

is given by the vector $v = (v_1, \ldots, v_n)$ such that $\Sigma_{j=1}^{n} a_{ij} v_j = b_i$, $i = 1, \ldots, m$, and also the vector $u - v$ is orthogonal to the subspace defined by the equations

$$\sum_{j=1}^{n} a_{ij} x_j = 0, \qquad i = 1, \ldots, m. \tag{1.3.11}$$

The orthogonality of $u - v$ to the subspace defined by (1.3.11) is equivalent to the condition that $u - v = \Sigma_{i=1}^{m} \beta_i a_i$, where the β_i are some numbers and $a_i = (a_{i1}, \ldots, a_{in})$ are vectors. If the vector u and (1.3.10) are given, the β_i are uniquely determined.

Indeed, let

$$v = u - \sum_{i=1}^{m} \beta_i a_i.$$

Then, from the definition of v it follows that the β_i must satisfy the equations

$$\sum_{j=1}^{n} a_{ij}u_j - \sum_{r=1}^{m} \left(\beta_r \sum_{j=1}^{n} a_{rj}a_{ij}\right) = b_i, \quad i = 1, \ldots, m. \qquad (1.3.12)$$

Therefore, if the system of vectors $a_1, \ldots, a_m$ has rank m, the β_i, $i = 1, \ldots, m$, are uniquely determined from (1.3.12).

One can show that for any system of equations (1.3.10) there exists a matrix P such that for any vector u its projection onto the set defined by (1.3.10) is the vector Pu. For a detailed treatment of the projection matrix P, see Gill and Murray (1974). Similar methods also exist for problems with nonlinear constraints. However, in this case the application becomes considerably more complex.

The following general statement of mathematical analysis shows how projection methods may be used in constrained optimization problems. The gradient of the restriction of the function F(x) onto the set (1.3.10) at a point x^1 is the projection of the gradient $\nabla F(x^1)$ from the unrestricted problem onto the set (1.3.10).

2

Decomposition Methods

2.1 NATURE OF DECOMPOSITION METHODS AND PROBLEMS IN CONSTRUCTING EFFECTIVE ALGORITHMS

Decomposition methods are designed for solving large-scale mathematical programming problems that can be reduced from finding the solution of the original problem to that of solving a sequence of problems which either have a special structure that is easier to solve or are of lower dimension than the original problem. Normally, decomposition methods are constructed so that the subproblems can be solved independently and simultaneously, and then iteratively coordinated. Because of this feature, decomposition algorithms are often interpreted as representing the coordination of local solutions of subsystems in a large-scale control system.

The general theory of decomposition methods is closely related to the general theory of mathematical programming and can be viewed as an offshoot of the latter. Decomposition algorithms are tantamount to specific applications of more general algorithms (e.g., the Dantzig-Wolfe algorithm is a special application of the simplex method, the Kornai-Liptak algorithm is a particular application of the Brown-Robinson algorithm for solving a game, etc.), and the subproblems formed by decomposing the original problem are usually standard mathematical programming problems.

It is no simple task to construct a decomposition algorithm that will converge to the solution of the original problem. For example, a seemingly natural way to decompose the problem

$$f(x) \to \max;$$
$$g_s(x) \leq 0, \qquad s \in S,$$

would be to use the group coordinate descent algorithm, that is, solve (1.3.1) by successively fixing the values of different groups of the variables. However, for fairly general sets of assumptions on the functions $f(x)$ and $g_s(x)$, this algorithm will not converge to the desired solution.

The goal of this chapter is not to present an exhaustive survey of decomposition methods. Rather, we describe only those methods that are related in some fashion to currently existing iterative aggregation algorithms.

We will discuss decomposition algorithms for the general convex programming problem of the following form:

$$f(x_1, x_2, \ldots, x_J) \to \max; \tag{2.1.1}$$

$$g_s(x_1, x_2, \ldots, x_J) \leq b_s, \qquad s \in S; \tag{2.1.2}$$

$$x_j \in Q_j, \qquad j = 1, \ldots, J, \tag{2.1.3}$$

where S is the set of indices of the constraints (2.1.2). The vector of variables x is partitioned into the subvectors $x_1, \ldots, x_J$, where the subvector x_j consists of the variables belonging to subsystem j; the functions g_s are the global constraints on the resources or objectives in the system; and the set Q_j $(j = 1, \ldots, J)$ describes the production possibilities of subsystem j (i.e., it implicitly describes the local constraints on the vector x_j independent of the rest of the system). The constraints in (2.1.2) are called the global constraints and the constraints in (2.1.3) are called the local constraints.

We assume that the objective function f and the global constraints g_s are separable with respect to x_j, namely,

$$f(x_1, \ldots, x_J) = \sum_{j=1}^{J} f_j(x_j); \tag{2.1.4}$$

$$g_s(x_1, \ldots, x_J) = \sum_{j=1}^{J} g_{sj}(x_j). \tag{2.1.5}$$

Thus problems to which we apply decomposition algorithms consist of three elements: the functional f to be maximized, the global constraints g_s, and the local constraints Q_j.

We adopt a formal classification system that separates decomposition methods into three groups. The first group of methods are based on the idea that the subsystems provide the center with information about their production possibilities (e.g., about the set Q_j), and this information reflects the actual production possibilities in a neighborhood of the optimal solution more and more precisely from iteration to iteration. The information provided by the subsystems is generated by solving a local optimization problem in response to a request from the center. Each of the local problems contains the local constraints (2.1.3), while the optimality criteria used differ from algorithm to algorithm. Examples of optimality criteria used are: the maximum value of the subsystems' activities computed from prices provided by the center; the minimal deviation from the solution provided by the center (not necessarily in Q_j); and so on. As the center accumulates these responses it combines them in an optimal manner to form new values of the appropriate parameters used by the subsystems for their particular optimality criteria. Thus in the first group of algorithms the center concentrates the detailed information about the entire system in a neighborhood of an optimal solution, and this information is provided by the subsystems. These algorithms are described in detail in Section 2.2.

The second and third groups of algorithms utilize the idea that if the global constraints (2.1.2) are eliminated, the resulting problem would break down into J independent subproblems. (By "elimination" of the global constraints we mean either their transformation into some other form or some indirect consideration of them.)

In the second group of algorithms the highest possible value b_{sj} is fixed for each component in (2.1.5), so that $\Sigma_j b_{sj} = b_s$, and the original problem is partitioned into J independent subproblems. The center then computes new values for the b_{sj} based on the solutions obtained from the local problems.

This approach can be interpreted as the center allocating global resources and objectives to the subsystems, while each subsystem solves its own local problem but includes also the additional global constraints and objectives passed to it from the center. Thus in the second group of algorithms the global constraints are transformed into local constraints by direct allocation of the global resources and objectives among the subsystems.

There is a special subgroup of algorithms that have features of both the first and second groups of decomposition algorithms. In this group of algorithms the production sets Q_j in (2.1.3) are not approximated; rather, the set of all possible consumptions of the global resources and of the feasible global objectives in each subsystem is approximated. In terms of their economic interpretation, these algorithms belong to the second group of algorithms. The algorithms that use direct allocation of resources are described in Section 2.3.

In the third group of decomposition algorithms, discussed in Section 2.4, the global constraints are considered only indirectly, by modifying the objective function. There are two approaches to appending the global constraints to the objective function. One is to use penalty functions and the other is to use Lagrangian functions. Direct application of the first approach does not allow us to break the original problem into smaller subproblems. To achieve this, the penalty function method is "built into" the algorithm with direct allocation of resources (i.e., in some algorithm of the second group). The resulting allocation of resources is more flexible here than in the algorithm of the second group since the subsystems are allowed to deviate to a certain degree from the allocation given by the center, but these deviations are penalized in the objective function. The advantages of such "elastic" allocations are discussed in Section 2.4.

The use of the classical Lagrangian function in a decomposition algorithm allows the original problem to be partitioned into independent subproblems. However, it is applicable only if the original problem is strictly convex, which is not true for linear programming problems, for example. Therefore, we restrict our attention to the so-called "modified Lagrangian functions," which can be viewed as a synthesis of penalty functions and classical Lagrangian functions.

In order that the modified Lagrangian function break the initial problem into smaller subproblems, we use a special modified Lagrangian function containing not only the primal and dual variables, but also the resources and the objectives allocated to the subsystems by the center. In this instance, as with penalty functions methods, the global constraints are appended to the objective function of the problem, allowing for a flexible allocation of the resources at each iteration.

2.2 ALGORITHMS THAT APPROXIMATE THE SET OF FEASIBLE SOLUTIONS OF THE LOCAL PROBLEMS

The first decomposition algorithm for solving linear programming problems with a block-structured constraint matrix was proposed by Dantzig and Wolfe (1960, 1961). A block-structured linear programming problem has two types of constraints: constraints on the general resources used by all the subsystems in the overall system and constraints on the local resources that are only used by a single subsystem. Formally, this problem has the following form:

$$\sum_k c_k \xi_k \to \max; \tag{2.2.1}$$

$$\sum_k r_{sk} \xi_k \leqq b_s, \qquad s \in S_0; \tag{2.2.2}$$

$$\sum_{k \in K_j} r_{sk} \xi_k \leqq b_s, \qquad s \in S_j, \qquad j = 1, \ldots, J; \tag{2.2.3}$$

$$\xi_k \geqq 0, \qquad k \in K_j, \qquad j = 1, \ldots, J, \tag{2.2.4}$$

where k is the technology index ($k = 1, \ldots, K$); j a subscript denoting the subsystem ($j = 1, \ldots, J$); s a subscript denoting a limited resource; K_j a set of technologies used by subsystem j; S_0 the set of global resources; S_j a set of local resources of subsystem j; ξ_k the unknown rate of utilization of technology k (i.e., the volume of output produced by the technology k; c_k the unit value of product, manufactured by technology k; b_s the availability of resource s; and r_{sk} the input of resource s per unit of product manufactured by technology k.

The constraint coefficient matrix R for this problem has the form

$$\begin{bmatrix} R_1 & R_2 & \cdots & R_J \\ R^1 & 0 & \cdots & 0 \\ 0 & R^2 & \cdots & 0 \\ \cdot\cdot & \cdot\cdot & \cdots & \cdot\cdot \\ 0 & 0 & \cdots & 0 \end{bmatrix}, \tag{2.2.5}$$

where R_J is a matrix of dimension $|S_0| \times |K_j|$ and R^J is a matrix of dimension $|S_j| \times |K_j|$, where $|S|$ denotes the cardinality of the set S. This implies that the coefficient matrix for this problem has all its nonzero elements contained either in submatrices (blocks) that have no common rows or columns or else in a submatrix that is not part of any block.

The main idea of the Dantzig-Wolfe algorithm is to use the block structure of the coefficient matrix of the original problem in order to solve the larger original problem by solving a sequence of smaller local subproblems (determined by the number of blocks) and a coordinating master problem. At each iteration of the algorithm the local problems are formed using the local constraints in a given block and the local linear optimality criterion. The variable coefficients in this optimality criterion are modified on the basis of the prices (dual variables) determined by the master problem during the previous iteration.

Suppose that at iteration $\sigma - 1$ the variables $\xi_k^{\sigma-1}$ from (2.2.1)-(2.2.4) have already been found. Using the values $\varepsilon_k^{\sigma-1}$, $k \in K_j$, that are relevant to subsystem j, this subsystem computes the amount of the global resources it utilizes from the formula

$$b_{sj}^{\sigma} = \sum_{k \in K_j} r_{sk}\xi_k^{\sigma-1}, \qquad s \in S_0, \tag{2.2.6}$$

and the combined value of products of subsystem j from the formula

$$c_j^\sigma = \sum_{k \in K_j} c_k \xi_k^{\sigma-1} \tag{2.2.7}$$

and passes these values to the master problem.

The master problem is formed from the global constraints and the linear objective function. The variables in the master program represent the relative weight z_j given the various input-output plans furnished by the subsystems from the previous iterations. Formally, the master problem is

$$\sum_{j=1}^{J} \sum_{\tau \in T_j^\sigma} c_j^\tau z_j \to \max; \tag{2.2.8}$$

$$\sum_j \sum_{\tau \in T_j^\sigma} z_j b_{sj}^\tau \leqq b_s, \qquad s \in S_0; \tag{2.2.9}$$

$$\sum_{\tau \in T_j^\sigma} z_j = 1, \qquad j = 1, \ldots, J, \qquad z_j \geqq 0, \tag{2.2.10}$$

where T_j^σ contains those iteration numbers τ that the plan generated by subsystem j included in the basis for the center's optimal plan at iteration $\sigma - 1$. This implies that all variants not included in the solution at iteration $\sigma - 1$ (i.e., have zero weight) are not considered at the next iteration.

Thus at each iteration only a subset of the plans generated by the subsystems from the previous iterations is considered in the center's problem at the next iteration. At any iteration, the number of variants contained in this subset cannot exceed $|S_0| + J$, where $|S_0|$ is the number of global constraints in the original problem.

Comparing the number of constraints (2.2.1)-(2.2.3) in the original problem with the number of constraints (2.2.9)-(2.2.10) in the master problem, we see that the global constraints (2.2.2) in the original problem have been transformed into the constraints (2.2.9) of the center's problem, while the local constraints (2.2.3) of the original problem have been transformed into the constraints (2.2.10);

hence at each iteration the center solves a problem with $|S_0| + J$ linear constraints, whereas the original problem contains $\Sigma_{i=0}^{j} |S_i|$ constraints. It can be formally demonstrated that the nature of the transformation to the constraints (2.2.9), (2.2.10) lies in the fact that the sets that satisfy the local constraints (2.2.3) are approximated in the center's problem at iteration σ by the polyhedron spanned by the vectors x_j^τ, $\tau \in T_j^\sigma$, where x_j^τ is the solution vector of subsystem j's local problem at iteration $\sigma - 1$, that is,

$$x_j^\tau = (\xi_k^\tau | k \in K_j).$$

The solution to the dual of the master problem (2.2.8)-(2.2.10) generates the prices p_s, $s \in S_0$, that correspond to the global constraints (2.2.9). These prices are passed to the subsystems in order to calculate the modified objective function coefficients for the local problems of the form

$$\hat{c}_k^\sigma = c_k - \sum_{s \in S_0} p_s^\sigma r_{sk}. \tag{2.2.11}$$

The subsystems then solve the appropriate local problem, which has the following form for subsystem j at iteration σ:

$$\sum_{k \in K_j} \hat{c}_k^\sigma \xi_k \to \max; \tag{2.2.12}$$

$$\sum_{k \in K_j} r_{sk} \xi_k \leqq b_s, \qquad s \in S_j; \tag{2.2.13}$$

$$\xi_k \geqq 0, \qquad k \in K_j. \tag{2.2.14}$$

From the solution ξ_k^σ, $k \in K_j$, of the local problem (2.2.12)-(2.2.14), each subsystem calculates the quantities $b_{sj}^{\sigma+1}$, $s \in S_0$, and the coefficients $c_j^{\sigma+1}$, which are then passed to the center. The center then solves the updated master problem and obtains the new prices $p_s^{\sigma+1}$ for the global resources from which each subsystem's present plan can be evaluated. The search for the solution of the original problem is terminated when a stopping criterion is satisfied. The solution to the original problem is found from the formula

$$\xi_k = \sum_{\tau \in T_j^\gamma} z_j^{\tau\gamma} \xi_k^\tau, \qquad k \in K_j, \qquad j = 1, \ldots, J, \tag{2.2.15}$$

where γ is the number of the last iteration and $z_j^{\tau\gamma}$ is the solution of the master problem obtained at iteration γ.

Mathematically, the algorithm above can be seen to be a direct application of the primal simplex method to an auxiliary problem generated from the original problem. Each iteration of the simplex method consists of the following stages. At the start of the first stage we have columns that form the current basis and one or more columns that are introduced into the basis. From this expanded set of columns we compute a new basis and we calculate the prices (dual variables) for the constraints. In the second stage we search for columns whose associated dual variables have positive (preferably the largest) values. If none of the dual variables are positive, the algorithm stops. Otherwise, the columns that have been found are incorporated into the basis at the first stage of the next iteration.

Since the Dantzig-Wolfe decomposition algorithm first appeared, a number of other decomposition algorithms for solving linear programs have employed an analogous scheme that involves passing from the subsystems to the center information on the variants of the consumption of the global resources and each subsystem's value of products, and the passing from the center to the subsystem's information about the value of the resources. These algorithms differ mainly in how the master problem is constructed and in the form of the objective function for the local problems. Mathematically, these algorithms differ from the Dantzig-Wolfe algorithm in that they use not only the primal simplex algorithm but also other well-known algorithms for solving linear programming problems. Thus Abadie and Williams (1963) use the dual simplex algorithm and Bell (1965) uses the Ford-Fulkerson error minimization algorithm. [A detailed description of these algorithms can be found in Lasdon (1970).]

In these algorithms it becomes more difficult to find the solution of the local problems, as the objective functions are linear-fractional functions. An attempt to overcome this difficulty was presented by Movshovich (1966) and Golstein (1966). If the original problem consists of block matrices with both coupling constraints (i.e., global resources) and common variables, direct application of the Dantzig-Wolfe algorithm would lead to a three-level algorithm. Some possible approaches for solving this problem have been proposed by Rosen (1964) and Ritter (1967).

The Dantzig-Wolfe decomposition algorithm has been extended to nonlinear convex programming problems by Malinvaud (1968) and Polterovich (1969). What distinguishes the convex case from the linear case is that variants rejected at a previous iteration may have to be considered at future iterations. Due to the nonlinearity of the constraints, we cannot be sure in advance that a solution to the local problem that has been tried previously will not be part of a linear combination that is optimum for the entire problem. As a result, these algorithms require an increasing amount of information about the solutions of both the master and subproblems to be stored from iteration to iteration, as well as an increasing number of variables in the master problem.

Decomposition algorithms for convex problems that are based on the Dantzig-Wolfe algorithm preserve the main features of this algorithm in the linear case. That is, the local constraints (2.2.3) are considered only directly in the subsystems' problems, while in the master problem they are approximated by a set of constraints with simple structure. There are several different decomposition algorithms for convex programming problems that differ from each other in the sets they use to approximate the local constraints and in the choice of objective functions in the local problems. The differing local objective functions correspond to different ways that the center can influence the subsystems in order to adjust their plans to be more in line with the preferences of the center.

We now describe in detail the convex programming problem to be decomposed, and then present two possible decomposition algorithms from Polterovich (1969) for solving this problem. As before, assume that the variables are partitioned into groups that correspond to subsystems (i.e., for any j the vector x_j only contains variables belonging to subsystem j). We assume that the set of variable subscripts $\{1,2,\ldots,K\}$ is partitioned into the disjoint subsets K_j:

$$\{1,2,\ldots,K\} = \bigcup_{j=1}^{J} K_j, \qquad K_i \cap K_j = \emptyset, \qquad i \neq j \text{ and}$$
$$x_j = (\xi_k | k \in K_j).$$

The convex programming problem with both local and global constraints is given by

$$f(x_1,\ldots,x_J) \to \max; \tag{2.2.16}$$

$$x = (x_1,\ldots,x_j) \in T; \tag{2.2.17}$$

$$x_j \in Q_j, \qquad j = 1, \ldots, J. \tag{2.2.18}$$

Here T denotes the set of all vectors x that satisfy the constraints

$$g_s(x) \leq b_s, \qquad s \in S, \tag{2.2.19}$$

where $g_s(x)$ are convex functions and $f(x)$ is a concave function of x. We assume that for each j the set Q_j is convex and closed.

Let Q be the Cartesian product of sets $\Pi_{j=1}^{J} Q_j$; that is, the set Q_j is composed of the set of vectors $x = (x_1,\ldots,x_J)$ satisfying $x_j \in Q_j$ for $j = 1, \ldots, J$. Then the constraints (2.1.17), (2.1.18) can be represented as

$$x \in T \cap Q. \tag{2.2.20}$$

In the first algorithm the sets defined by the local constraints are represented by an "inner approximation;" that is, at each iteration σ the set Q_j will be approximated in the master program by the set Q_j^σ, which is contained in Q_j. As the iteration number σ increases the approximating sets, the Q_j^σ expand so as to approximate Q_j in a neighborhood of the optimum with ever-increasing precision. In the second algorithm the sets Q_j are represented by an "outer approximation;" that is, for all iterations σ the sets Q_j^σ contain the sets Q_j. When progressing from iteration σ to iteration $\sigma + 1$

the set $Q_j^{\sigma+1}$ is obtained from the set Q_j^σ by deleting a certain subset of Q_j^σ not being in Q_j. In short, in both algorithms, as the number of iterations σ increases, the sets Q_j^σ will approximate the sets Q_j with ever-increasing precision in a neighborhood of the optimal solution.

In more detail, in the first algorithm at iteration σ the center solves the following master problem:

$$\begin{aligned} & f(x) \to \max; \\ & x \in T; \\ & x_j \in Q_j^{\sigma-1}, \end{aligned} \tag{2.2.21}$$

where $Q_j^{\sigma-1}$ is the convex hull of the vectors $x_j^1, \ldots, x_j^{\sigma-1}$; that is,

$$Q_j^{\sigma-1} = \left\{ \sum_{\tau=0}^{\sigma-1} \lambda_j^\tau x_j^\tau \,\middle|\, \sum_{\tau=0}^{\sigma-1} \lambda_j^\tau = 1,\ \lambda_j^\tau \geqq 0 \right\}.$$

In solving this problem the center finds the solution $\tilde{x}^\sigma$ and the vector p^σ that is normal to the supporting hyperplane of the set $T \cap Q^{\sigma-1}$ at the point $\tilde{x}^\sigma$ (this vector forms an obtuse angle with all vectors of the form $x - \tilde{x}^\sigma$, $x \in Q^{\sigma-1}$, i.e., $(p^\sigma, x - \tilde{x}^\sigma) \leqq 0$ and an acute angle with all vectors $x - \tilde{x}^\sigma$, $x \in L^\sigma = \{x \in T | f(x) \geqq f(\tilde{x}^\sigma)\}$).

The vector p^σ is partitioned into p_j^σ subvectors that consist of the components corresponding to subsystem j.

Subsystem j solves the problem

$$(p_j^\sigma, x_j) \to \max; \qquad x_j \in Q_j, \tag{2.2.22}$$

from which we derive the components of the vector x^σ that corresponds to subsystem j.

The algorithm works as follows. Suppose that $x_j^0, \ldots, x_j^{\sigma-1} \in Q$, are given. Then the polyhedron $Q_j^{\sigma-1}$ spanned by x_j^τ, $0 \leqq \tau < \sigma$, is an approximation of Q_j and satisfies $Q_j^{\sigma-1} \subset Q_j$. A necessary and sufficient condition for (2.2.21) to have a solution for all σ is that $x_0 \in T \cap Q$, since this implies that $T \cap Q^{\sigma-1}$ is not empty for all σ.

Having solved (2.2.21), it is desirable to construct the local problems so that they will expand the sets $Q_j^{\sigma-1}$ such that $f(x^{\sigma+1})$ will be greater than $f(x^\sigma)$ unless x^σ is the solution to the original problem. In addition, the local problems should reveal whether x^σ is the solution to the entire problem or the process needs to continue for another iteration. A possible approach to solving these problems is to find a vector p^σ which points from $Q^{\sigma-1}$ in the direction of increase for $f(x)$. According to the theory of separating hyperplanes, such a vector must exist. For definiteness we can normalize p^σ to satisfy $\|p^\sigma\| = 1$. Now the set $Q^{\sigma-1}$ can be expanded by adding to it the vector x^σ which lies in Q and which has maximal projection onto p^σ among all vectors $x \in Q$, that is,

$$(p^\sigma, x) \to \max; \qquad x \in Q. \tag{2.2.23}$$

This problem is equivalent to solving the sequence of problems

$$\begin{aligned} &(p_j^\sigma, x_j) \to \max; \\ &x_j \in Q_j, \qquad j = 1, \ldots, J. \end{aligned} \tag{2.2.24}$$

It can be shown that if x^σ is contained in $Q^{\sigma-1}$ (i.e., $Q^{\sigma-1} = Q^\sigma$), then $\tilde{x}^\sigma$ is the solution to the entire problem.

The condition $(p^\sigma, x - x^\sigma) \leqq 0$, $x \in L^\sigma$, ensures that x^σ expands $Q^{\sigma-1}$ in the desired direction, for otherwise Q^σ may better approximate Q than does $Q^{\sigma-1}$ but lie farther from x^*.

As the problem is convex, we can always take p^σ equal to

$$\nabla f(x^\sigma) - \Sigma_s \lambda_s \nabla g_s(x^\sigma), \tag{2.2.25}$$

where λ_s is the solution to the dual of the center's problem (2.2.21), that is, the gradient vector with respect to x of the Lagrangian function at the solution point of (2.2.21). When p^σ is chosen in this fashion the algorithm has the following interpretation. Let $Q^{\sigma-1}$ be given. Then we search over $Q^{\sigma-1}$ for the saddle point of the Lagrangian function

$$F(x,\lambda) = f(x) - (\lambda, g(x)), \qquad x \in Q^{\sigma-1}, \qquad \lambda \geqq 0,$$

of the problem

$$f(x) \to \max; \qquad x \in T, \tag{2.2.26}$$

or equivalently we maximize the function

$$\varphi(x) = \min_{\lambda \geq 0} F(x,\lambda)$$

on the set $x \in Q^{\sigma-1}$. At the solution, if the gradient $\nabla_x F(x,\lambda)$ is directed away from the set $Q^{\sigma-1}$, then $Q^{\sigma-1}$ is expanded in this direction. Here the symbol ∇ may also mean the generalized gradient.

Polterovich (1968) has shown that if $\{T\} \cap Q \neq \emptyset$ and is bounded (where $\{T\}$ denotes the interior of T), the algorithm converges. We can see that in this algorithm the sets Q are approximated from the inside. This implies that the sequence of solution vectors $\tilde{x}^{\sigma}$, $\sigma = 0, 1, \ldots$, are all feasible in the original problem.

The second algorithm, which uses an outer approximation to Q, consists of the following steps. We are given the vectors $x_j^0, \ldots, x_j^{\sigma-1}$ in Q_j, $j = 1, \ldots, J$, as well as $\tilde{x}^0, \ldots, \tilde{x}^{\sigma-1}$ in T. The center solves the problem

$$f(x) \to \max; \qquad x_j \in Q_j^{\sigma-1}; \qquad x \in T, \tag{2.2.27}$$

where the set $Q^{\sigma-1}$ is given by

$$Q_j^{\sigma-1} = \bigcap_{\tau=0}^{\sigma-1} V_j^{\tau}; \qquad V_j^{\tau} = \{x_j \,|\, (x_j - x_j^{\tau}, \tilde{x}_j^{\tau} - x_j^{\tau}) \leqq 0\}.$$

The solution $\tilde{x}^{\sigma}$ to (2.2.27) is partitioned into subvectors $\tilde{x}_j^{\sigma}$ ($\tilde{x}_j^{\sigma}$ contains the components of $\tilde{x}^{\sigma}$ corresponding to subsystem j), which are then passed to the subsystems. With this information subsystem j finds the vector x_j^{σ} by solving the problem

$$\|x_j - \tilde{x}_j^{\sigma}\| \to \min; \qquad x_j \in Q_j. \tag{2.2.28}$$

The new approximation x^{σ} is formed by combining the solutions x_j^{σ} generated by the subsystems. If, at some iteration σ, x^{σ} should lie in Q, then x^{σ} would be the solution to the original problem.

A more detailed description of this algorithm is as follows. At iteration σ the set $Q^{\sigma-1}$ is constructed as the intersection of the half-spaces V^τ, $\tau = 0, 1, \ldots, \sigma - 1$, where the half-spaces V^τ are defined as $V^\tau = \{x | (x - x^\tau, \tilde{x}^\tau - x^\tau) \leqq 0\}$. The half-spaces V^τ can be viewed as eliminating extraneous parts from the set T.

The center's problem then consists of solving (2.2.27) with $Q^{\sigma-1}$ defined as above. If the solution to the center's problem $\tilde{x}^\sigma$ lies in Q, then $\tilde{x}^\sigma$ is the solution to the entire problem (2.2.16)-(2.2.18). If, however, $\tilde{x}^\sigma \notin Q$, we need to construct a linear manifold that separates $\tilde{x}^\sigma$ from Q. This separating linear manifold can be found by finding a point in Q that is closest to $\tilde{x}^\sigma$, that is, the solution to the problem

$$\|x - \tilde{x}^\sigma\|^2 \to \min; \qquad x \in Q, \tag{2.2.29}$$

and take V^σ as the set $\{x | (x - x^\sigma, \tilde{x}^\sigma - x^\sigma) \leqq 0\}$, where x^σ is the solution of (2.2.29). Solving the minimization problem (2.2.29) is equivalent to solving the sequence of problems

$$\|x_j - \tilde{x}_j^\sigma\| \to \min; \qquad x_j \in Q_j, \qquad j = 1, \ldots, J, \tag{2.2.30}$$

and hence can be solved at the subsystem level.

Polterovich (1969) has proven that if the function f is continuous, the set T closed, the set Q closed and convex, and the set $L = \{x | x \in T, f(x) \geqq f(x^*)\}$ is bounded [here x^* is the solution of (2.2.16)-(2.2.18)], then any limit point of the sequence $x^0, \ldots, x^\sigma$ lies in Q and $f(x^\sigma)$ converges to $f(x^*)$ as σ goes to infinity.

Finally, note that for linear programming problems the inner and outer approximation problems are dual to each other in the sense that the solution of the primal problem by one of the methods is equivalent to the solution of the dual problem by the other method.

These types of algorithms have only been proposed for linear and convex programming algorithms, as it is much more difficult to approximate nonconvex sets; however, the outer approximation algorithm can be used to solve problems where the objective function f(x) is nonconvex as long as the constraints are convex. In linear problems the set of feasible points determined by the local constraints can be represented by a polyhedron, that is, by the convex hull of a finite number of vertices and infinite edges. As a result, the algorithm is finite for linear problems. For the general convex problem it is not always possible to find a finite number of points and vectors whose convex hull coincides with the set defined by the local constraints.

Furthermore, since for nondegenerate solutions of linear programming problems the solution vector contains only as many nonzero elements as there are essential constraints in the problem, the amount of information required at the center level increases in this case only up to a certain point. Since this is not necessarily true in the convex case, the information needed at the center level is constantly increasing; this in turn implies that the center must solve a sequence of problems of increasing complexity. Also, in the Dantzig-Wolfe algorithm all global constraints are retained in the master problem at each iteration. Therefore, if there are a large number of global constraints, the decomposed master problem may be almost as difficult to solve as the original problem.

There is also one other problem with a Dantzig-Wolfe type of algorithm, which can be illustrated for the case of a nondegenerate linear programming problem (i.e., having unique primal and dual optimums). It is well known that for such a problem the number of nonzero components in the optimal solution vector is equal to the number of linearly independent essential constraints at the optimal solution (an essential constraint is one that becomes an equality at the optimum point). Let n_0 denote the number of essential global constraints and n_j the number of essential constraints in block j. Therefore, for a nondegenerate problem the total number of binding constraints at an optimum solution is $n_0 + \Sigma_j n_j$. If all the local problems have nondegenerate optimal solutions, the total number of nonzero components in their solution is $\Sigma_j n_j$. This means that at the solution to the original problem, some of the local problems do not have unique solutions, and the sum of the dimensions of the polyhedrons defined by the solutions to the local problems (recall that an optimal solution to a linear problem defines a polyhedron) is equal to n_0, the number of essential global constraints. For each of these polyhedrons the solution of the subproblem corresponding to the (unique) optimal solution of the entire problem can be found as the convex combination of its vertices. The iterative process terminates only after the center has accumulated information about a sufficient number of these vertices. Thus even at the optimal solution it is necessary to perform several iterations of the algorithm to ensure that the point indeed is the optimal point. One can see that the number of such iterations cannot be less than n_0/J.

It should be noted that global convergence to the optimal solution has been proven for the class of decomposition methods discussed above. Lasdon (1970) has pointed out that this class of decomposition methods has been applied successfully to solving a number of optimal problems, in particular problems with so-called "narrow blocks" (or generalized upper bounds), where each block has only one constraint, as long as advantage is taken of the special structure of the problems.

2.3 ALGORITHMS OF DIRECT DISTRIBUTION OF RESOURCES THAT CONVERT GLOBAL CONSTRAINTS INTO LOCAL CONSTRAINTS AT EACH ITERATION

To analyze this group of algorithms, we use as an example the problem (2.2.16)-(2.2.19) with the objective function f and the constraints g_s, $s \in S$, separable in j, that is, of the form $f(x) = \Sigma_j f_j(x_j)$, $g_s(x) = \Sigma_j g_{sj}(x_j)$ for $s \in S$. This type of algorithm has been generalized to nonseparable problems also; some general approaches have been discussed by Levin and Tanaev (1974). However, we will only consider the separable case so as not to complicate the discussion with nonessential details.

When the global constraints in the original problem represent constraints on general resources, the algorithm can be interpreted as follows. If the amount of the general resource allocated to each subsystem j (j = 1, ..., J) is fixed, the problem naturally decomposes into J independent subproblems, one for each subsystem. Let $F_j(B_j)$ denote the optimal value of the objective function of subsystem j, where the vector $B_j = (b_{sj} | s \in S)$ denotes the fixed allocation of the general resources. Then the sum of the optimal solutions of the subproblems, that is, $F(B) = \Sigma_j F_j(B_j)$, where the vector $B = \{b_{sj} | s \in S, j = 1, \ldots, J\}$, shows the effect of a given allocation of the resources among the subsystems.

Let us examine how this value varies with changes in the resource allocation among the subsystems [we shall denote this function by F(B), which is the master objective function]. The center's problem consists of optimizing this function subject to constraints that define the set of feasible allocations between the subsystems. Thus the solution of the original problem reduces to searching for an (optimal) allocation of the general resources between the subsystems that would maximize the master objective function. Algorithms that utilize this approach differ only in the method used for seeking the maximum of the master objective function.

Formally, algorithms of this type are constructed around the observation that the problem

$$\max \Sigma_j f_j(x_j);$$
$$\Sigma_j g_{sj}(x_j) \leqq b_s, \qquad s \in S, \tag{2.3.1}$$
$$x_j \in Q_j, \qquad j = 1, \ldots, J,$$

is equivalent to the problem

$$\max \Sigma_j f_j(x_j); \tag{2.3.2}$$
$$g_{sj}(x_j) \leqq b_{sj}, \qquad s \in S, \qquad j = 1, \ldots, J; \tag{2.3.3}$$
$$\Sigma_j b_{sj} = b_s, \qquad s \in S; \tag{2.3.4}$$
$$x_j \in Q_j, \qquad j = 1, \ldots, J. \tag{2.3.5}$$

Therefore, one may want to select values of b_{sj} that satisfy equation (2.3.4) and then solve equations (2.3.2), (2.3.3), and (2.3.5) with these values fixed (this causes the problem to be solved as J independent subproblems). Then one can compute the new values of b_{sj} that satisfy (2.3.4), and so on. However, unless special measures are taken, the problem defined by (2.3.2), (2.3.3), and (2.3.5) will have a degenerate solution in a neighborhood of the solution of the original problem.

The algorithms that use this method to solve (2.3.1) are based on the theory of marginal values for convex problems. This theory states that the value of a dual variable of inequality s is equal to $\partial F/\partial b_{sj}$. However, when $\partial F/\partial b_{sj}$ is used to recompute the solution of (2.3.4) to find an improved allocation, convergence is very slow. As with the Dantzig-Wolfe algorithm, the local problems for these algorithms are degenerate at the optimal point. Moreover, whereas in the Dantzig-Wolfe algorithm the total number of nonzero components in the solutions of the local problems is less than the number of nonzero components at the optimal solution, in the present algorithm the reverse is true. Hence the subsystems' problems are degenerate and the dual variables cease to convey useful information.

The problem (2.3.2)-(2.3.5) can be solved by finding the saddle point of its Lagrangian function. The computational process itself is organized in such a way that the values of the original variables are determined in the appropriate local problems. The saddle point can be found either by the method of fictitious games proposed by Brown (1951) and Robinson (1951) [and which was used in the Kornai-Liptak algorithm (1965)] or by the algorithm proposed by Arrow, Hurwicz, and Uzawa (1958), which is similar. Since in both of these algorithms the step size tends to zero, we cannot expect rapid convergence. Also, these algorithms may have the additional difficulty that the subproblems may not be solvable.

The first algorithm of this type was proposed by Kornai and Liptak (1965). At each iteration of this algorithm each subsystem optimizes the plan subject both to constraints on the amount of general resources they receive from the center and to constraints on the local resources if the original problem has a block structure [in their initial paper local constraints were not considered; that is, in (2.3.5) the set Q_j coincides with $\mathbb{R}^n_+$]. The master problem obtains from each subsystem the changes to the output and consumption of general resources in the plan as well as the value these items have in the local problems. On the basis of this information the master problem reallocates the general resources to the individual subsystems. The general resources are reallocated such that a subsystem that places a higher value on a given resource is given an increased share of that resource, while a subsystem that places a lower value on the resource is given a smaller share of the resource.

At each iteration the master objective function is approximated by a linear function constructed from the dual variables passed from the subsystems. Since this implies that the master problem is unconstrained, additional restrictions from above are imposed on the resource allocation and, to ensure convergence, solutions from successive iterations are averaged.

The Kornai-Liptak algorithm has been generalized to convex problems by Volkonsky (1965a) and Polterovich (1969). Assume that the present approximation to the optimal primal variables and to the dual-price vector of the resources $\lambda_j^{\sigma-1} = (\lambda_{sj}^{\sigma-1} | s \in S)$ have been set by the subsystems. The center attempts to alter the plan of each subsystem by reallocating the resources and by altering the objective function of each subsystem. To do this the center solves the problem

$$\max\{\Sigma_j (\lambda_j^{\sigma-1}, B_j)\}; \tag{2.3.6}$$

$$\Sigma_j b_{sj} = b_s; \tag{2.3.7}$$

$$B \in H, \tag{2.3.8}$$

where

$$B = (b_{sj} | s \in S,\ j = 1, \ldots, J); \qquad B_j = (b_{sj} | s \in S).$$

The set H is the set of all vectors B for which the local problems have solutions. The set H is the direct product of the sets H_j defined as

$$H_j = \{B_j | \text{the constraints (2.3.3), (2.3.5) have a common solution for a given } j\}.$$

From the solution of (2.3.6)-(2.3.8) we obtain the vector $B^{\sigma} = (b^{\sigma}_{sj})$, which furnishes the new resource allocation between the subsystems.

Next, each subsystem solves the local problem

$$\max f_j(x_j); \tag{2.3.9}$$

$$g_{sj}(x_j) \leq b^{\sigma}_{sj}; \tag{2.3.10}$$

$$x_j \in Q_j. \tag{2.3.11}$$

Let x^{σ}_j denote the solution of (2.3.9)-(2.3.11) and λ^{σ}_{sj} the value of the dual variables that correspond to the constraints (2.3.10). This algorithm can be interpreted as a convex two-person zero-sum game. The payoff function for the game is given by

$$L(x,\lambda,b) = \sum_j f_j(x_j) - \sum_{s\in S} \sum_j \lambda_{sj}(g_{sj}(x_j) - b_{sj})$$

with the following constraints on x, λ, b:

$$x \in Q;$$

$$\Sigma_j b_{sj} = b_s, \qquad s \in S;$$

$$B \in H;$$

$$\lambda_{sj} \geqq 0.$$

The game proceeds by having the first player minimize the function L by choosing the vector λ, and the other player maximizes the function L by choosing x, b.

The game is equivalent to a two-person zero-sum game that has the payoff function

$$\varphi(\lambda,b) = \max_{x\in Q} L(x,\lambda,b).$$

In the game the player who maximizes the function L by controlling vector b can be viewed as the center that allocates resources between subsystems, and the player who wants to minimize the function by controlling λ can be viewed as the subsystems that determine the (marginal) values of the resources in each subsystem.

It is well known that for an arbitrary game if the players choose as their strategies the optimal responses to their opponents' strategies the process may happen to be nonconvergent. Robinson(1951) demonstrated that to ensure convergence of the process, it is sufficient to choose as one's strategy the generalized average over all of one's previous responses; that is,

$$B^\sigma = (1 - \gamma^\sigma)B^{\sigma-1} + \gamma^\sigma B^\sigma; \tag{2.3.12}$$

$$x_j^\sigma = (1 - \gamma^\sigma)x_j^{\sigma-1} + \gamma^\sigma x_j^\sigma; \tag{2.3.13}$$

$$\lambda_j^\sigma = (1 - \gamma^\sigma)\lambda_j^{\sigma-1} + \gamma^\delta \lambda_j^\sigma, \tag{2.3.14}$$

where

$$\gamma^\sigma \to 0, \qquad \sum_{\sigma=1}^{\infty} \gamma^\sigma = \infty. \tag{2.3.15}$$

Besides the above-mentioned procedures that iteratively recalculate x, b, and λ, we also must make certain that the subsystems' problems are solvable. This concern makes using this class of processes much more difficult. In the general case, rigorous methods that include (2.3.8) are cumbersome to compute. However, for a fairly common case, when we solve the linear program

$$\Sigma_j (c_j, x_j) \to \max;$$

$$\Sigma_j (a_{sj}, x_j) \leq b_s, \qquad s \in S;$$

$$x_j \in Q_j,$$

where all the components of the vectors a and b are nonnegative and where the sets Q_j are bounded and contain the zero vector, a sufficient condition for the local problems to be solvable is that $b_{sj} \geqq 0$ for all s, j. In other words, all the local problems are solvable if the constraint (2.3.8) is replaced by the condition $b \in \mathbb{R}_+^{|S|+J}$. The proof of convergence of this algorithm for the general case is given in Belenky and Volkonsky (1974).

Volkonsky (1965b) has proposed the following decomposition algorithm for the convex programming problem.[1] Assume as given the

---

[1]Polterovich (1970) has proven convergence of this algorithm for $g_{sj}(x_j) \geqq 0$, $g_{sj}(0) = 0$.

resource allocation $b_{sj}^{\sigma-1}$ satisfying (2.3.8) and assume also that (2.3.9)-(2.3.11) is solved using these values, and let λ_j^σ be the corresponding dual solution for block j. As has been noted,

$$\lambda_{sj} = \frac{\partial F_j(B_j)}{\partial b_{sj}},$$

and therefore it is desirable for the center to shift the present resource allocation to the allocation $\tilde{b}_{sj}^\sigma = b_{sj}^{\sigma-1} + \gamma^\sigma \lambda_{sj}^\sigma$, where γ^σ is a step size in the direction of the vector (λ_{sj}^σ). However, the change in the resource allocation caused by this step may mean that (2.3.8) is no longer satisfied; therefore, the center should choose a feasible vector that is located as close as possible to the vectors $\tilde{B}_j^\sigma$. In other words, the center should solve the problem

$$\sum_j \|B_j^{\sigma-1} + \gamma^\sigma \lambda_j^\sigma - B_j\| \to \min; \tag{2.3.16}$$

$$\sum_j b_{sj} = b_j, \qquad s \in S; \tag{2.3.17}$$

$$B \in H. \tag{2.3.18}$$

If in the problem (2.3.16)-(2.3.18) it is not possible to give a constructive description of the set H, we can attempt to find a subset of H that can be described more easily and which coincides with H in a neighborhood of the optimal solution. However, it is not always possible to find such a subset.

Martines-Soler and Cherniak (1974) have proposed another approach for handling incompatibility among the local subproblems. In this approach, the center solves the problem (2.3.16), (2.3.17) without considering the constraints (2.3.18), that is, by ignoring the set H. If all the local subproblems are compatible and the dual variables in (2.3.10) exist, these solutions can be used to continue the iterations. Otherwise, we have to find a direction M^σ satisfying

$$(M^\sigma, B - B^{\sigma-1}) \geqq 0 \qquad \text{for all } B \in H. \tag{2.3.19}$$

The vector M^σ calculated this way points "inside" the set H.

Further, instead of taking a step according to (2.3.12)-(2.3.14) in the direction M^σ, take

$$B^\sigma = B^{\sigma-1} + \gamma^\sigma M^\sigma \tag{2.3.12'}$$

and the vector $\lambda^{\sigma-1}$ does not change; that is, $\lambda^\sigma = \lambda^{\sigma-1}$.

This procedure can be implemented by using the following method to calculate the vector M^σ (see Dudkin, ed., 1979, Sec. 1.3). As the set H is the direct product of the subspaces H_j and since $(M^\sigma, b - b^\sigma) = \Sigma_j (M_j^\sigma, b_j - b_j^\sigma)$, to find a vector M^σ that satisfies (2.3.19) it is sufficient to solve a problem analogous to (2.3.19) for each subsystem j. More precisely, if $B_j^\sigma \notin \text{int } (H_j)$, set $M_j^\sigma = 0$, where int (H_j) is the set of interior points of the convex set H_j. Otherwise, we can find a λ_j^σ that satisfies the condition

$$(\lambda_j^\sigma, B_j - B_j^{\sigma-1}) \geqq 0 \qquad \text{for any } B_j \in H_j. \tag{2.3.20}$$

Since $B_j \in \text{int } (H_j)$ and H_j is convex, such a λ_j^σ should exist.

The vector λ_j^σ that solves (2.3.20) can be found by solving the maximim problem

$$\max_{\lambda_j} \min_{B_j} (\lambda_j, B_j - B_j^{\sigma-1}) \tag{2.3.21}$$

with the unknowns λ_j and B_j subject to the constraints

$$|\lambda_{sj}| \leqq 1, \qquad s \in S; \tag{2.3.22}$$

$$g_{sj}(x_j) \leqq b_{sj}, \qquad s \in S, \tag{2.3.23}$$

satisfied for some (at least one) set of x_j;

$$x_j \in Q_j. \tag{2.3.24}$$

[Note that in the reference given above there is a mistake in the description of the problem (2.3.21)-(2.3.24)—the minimization over the vector b_j was omitted in (2.3.21).]

When (2.3.16), (2.3.17) are solved without the constraint (2.3.18), the problem decomposes into solving |S| problems, one for each resource $s \in S$. These problems have explicit solution

$$\tilde{b}_s^{\sigma} = b_s^{\sigma-1} + \gamma^{\sigma}\lambda_s^{\sigma} - \frac{\gamma}{J}\sum_j \lambda_{sj}^{\sigma}. \tag{2.3.25}$$

There exists a class of algorithms whose economic interpretation is that of resource allocation, but in terms of their mathematical structure they solve the problem by approximating the domain of feasible solutions. However, in contrast to the algorithms in Section 2.2, they do not approximate the variables x_j of the original problem but instead approximate the feasible resource allocations b_{sj}; that is, they approximate the set H, or more precisely, its direct components H_j. These algorithms occupy an intermediate position between the algorithms of this section and those of Section 2.2. The most complete and explicit account of these algorithms is given in Weitzman (1970) and Silverman (1972). In the first algorithm the set of feasible resource allocations H_j is approximated from the outside by a decreasing sequence of polyhedrons, while in the second algorithm an inner approximation is used with an increasing sequence of polyhedrons. The first algorithm has local problems identical in form to (2.2.22) and its solution generates a new separating hyperplane (i.e., a new edge of the polyhedron). In the second algorithm the subsystems solve parametric problems on the basis of the direction of resource allocation furnished by the center; from this they find a new resource allocation (which is always feasible) that generates a new point for approximating the sets H_j $(j = 1, \ldots, J)$.

2.4 ELASTIC RESOURCE ALLOCATION ALGORITHMS WITH GLOBAL CONSTRAINTS APPENDED TO THE OBJECTIVE FUNCTION

There exists one other way of constructing decomposition algorithms for solving the problem

$$\sum_j f_j(x_j) \to \max; \tag{2.4.1}$$

$$\sum_j g_{sj}(x_j) \leq b_s, \qquad s \in S; \tag{2.4.2}$$

$$x_j \in Q_j, \tag{2.4.3}$$

where f_j is a concave function and g_{sj} is a convex function for all $s \in S$ and $j = 1, \ldots, J$ and Q_j is a closed convex set. In this method well-known nondecompositional methods for solving (2.4.1)-(2.4.3) are transformed in a manner such that the computation at each iteration can be broken down into a part performed by the center and a part performed by the subsystems. The two methods transformed in this manner are the penalty function method and the method of modified Lagrangian functions. This class of algorithms does not possess many of the undesirable computational properties of the resource-directed algorithms described above. If the original problem is solvable, so are all the subproblems, and since the dual prices generated from the subsystems are not used to redefine a resource allocation, the possible nonuniqueness of these solutions does not affect this class of algorithms.

The following decomposition algorithm, which uses the penalty function algorithm, is based on the ideas of Razumikhin (1967, 1975). The problem given by (2.4.1)-(2.4.3) can be reduced to the form of (2.3.2)-(2.3.5), which we repeat here.

$$\sum_j f_j(x_j) \to \max; \tag{2.4.4}$$

$$g_{sj}(x_j) \leq b_{sj}, \qquad s \in S, \qquad j = 1, \ldots, J; \tag{2.4.5}$$

$$x_j \in Q_j; \tag{2.4.6}$$

$$\sum_j b_{sj} = b_s, \qquad s \in S, \tag{2.4.7}$$

with auxiliary variables b_{sj}, $s \in S$, $j = 1, \ldots, J$. The penalty function algorithm solves (2.4.4)-(2.4.7) by imposing a penalty for violating the constraints (2.4.5). That is, the algorithm solves a sequence of problems of the form

$$\sum_j \left[f_j(x_j) - \sum_{s \in S} u_s(q_{sj}, g_{sj}(x_j) - b_{sj}) \right]; \tag{2.4.8}$$

$$x_j \in Q_j; \tag{2.4.9}$$

$$\sum_j b_{sj} = b_s, \qquad s \in S, \tag{2.4.10}$$

such that the sequence of numbers q_{sj} tends to infinity [$u_s(q,g)$ are penalty functions]. If as is usually the case, we assume that the penalty function is a quadratic function, (2.4.8) becomes

$$\sum_j \{f_j(x_j) - \sum_{s \in S} q_{sj}[(g_{sj}(x_j) - b_{sj})_{(+)}]^2\}. \qquad (2.4.11)$$

The computations proceed by solving (2.4.8)-(2.4.10) for fixed values of q_{sj} and then by recalculating of q_{sj} if the solution obtained is too infeasible. Therefore, the solution of (2.4.1)-(2.4.3) essentially reduces to the problem of solving (2.4.8)-(2.4.10), which, in turn, as in the algorithms of Section 2.3, is based on successively recalculating the b_{sj}'s and then solving (2.4.8)-(2.4.10) for fixed b_{sj}. Also, if at step σ all b_{sj} are fixed, then solving (2.4.8)-(2.4.10) decomposes into solving J subproblems of the form

$$\max \left\{f_j(x_j) - \sum_{s \in S} u_s(q_{sj}, g_{sj}(x_j) - b_{sj}^{\sigma-1})^2\right\}; \qquad (2.4.12)$$

$$x_j \in Q_j, \qquad (2.4.13)$$

and these problems can be solved at the subsystem level.

Thus the decomposition algorithm proceeds by having the J subsystems solve a problem of the form (2.4.12)-(2.4.13) and then having the center recalculate the quantities b_{sj}. The center derives the new values of the b_{sj} by solving the problem

$$\min \sum_s u_s(q_{sj}, g_{sj}(x_j^\sigma) - b_{sj}); \qquad (2.4.14)$$

$$\sum_j b_{sj} = b_s. \qquad (2.4.15)$$

This algorithm can be seen to be equivalent to solving (2.4.8)-(2.4.10) by group coordinate descent with respect to two groups of variables, x_j and b_{sj}. Since the constraints (2.4.9) and (2.4.10) each contain only one of the groups of variables, the coordinate descent algorithm will converge to the optimum. The penalty function decomposition algorithm originally proposed by Razumikhin (1967, 1975) was for linear problems of the form (2.4.1)-(2.4.3) and used a more general penalty function than the quadratic penalty function. It also uses a different procedure, based on physical analogies, for recalculating the b_{sj} at each iteration, which also ensures convergence of the algorithm. In this method the b_{sj} eventually can be expressed in terms of x^σ.

Finally, if in the penalty function decomposition algorithm the penalty function u_s are quadratic functions of $g_{sj}(x) - b_{sj}$, the solution of (2.4.14)-(2.4.15) can be found explicitly. If $\Sigma_j g_{sj}(x_j^\sigma) \leqq b_s$, we take for b_{sj}^σ the value $g_{sj}(x_j^\sigma) + \alpha_{sj}$, where α_{sj} are arbitrary nonnegative numbers that satisfy the condition $\Sigma_j \alpha_{sj} = b_s - \Sigma_j g_{sj}(x_j^\sigma)$. If $\Sigma_j g_{sj}(x_j^\sigma) > b_s$, the solution of (2.4.14)-(2.4.15) is given by

$$b_{sj}^\sigma = g_{sj}(x_j^\sigma) - \frac{1}{q_{sj}} \sum_s \frac{1}{q_{sj}} [b_s - \Sigma_j g_{sj}(x_j^\sigma)]. \qquad (2.14.16)$$

Based on this, the penalty function decomposition algorithm has the economic interpretation of solving (2.4.1)-(2.4.3) by having the center influence the subsystems by having it set a recommended level of resource consumption and setting a cost q_{sj} for violating this recommendation. In contrast to the algorithms of Section 2.3, the subsystems can violate the resource allocation recommended by the center by paying a penalty for this violation. It is in this sense that we term the resource allocation "elastic."

In real-life economic systems the plan-coordination process invariably includes prices. Therefore, the decomposition algorithms based on the modified Lagrangian function are of great economic interest. These algorithms include those proposed by Bakhmetyev et al. (1972). Telle (1975), and Razumikhin (1975, p. 180) (the algorithm for linear problems with a "shift of the constraints"). These algorithms solve (2.4.4)-(2.4.7) in a manner very similar to the algorithms based on penalty functions, except that the dual prices must be identical for each constraint in (2.4.5) indexed by the same s. However, although the algorithms based on modified Lagrangian functions are similar to those based on penalty functions, there are important differences between them. While the penalty function algorithms are equivalent to the gradient method with step size fixed for maximizing concave functions on a convex set, the algorithm of Telle (1975) is equivalent to the gradient method with fixed step size in order to find a saddle point. These three algorithms are all similar.

The first of these algorithms to appear was the one proposed by Bakhmetyev et al. for solving the problem

$$\sum_j f_j(x_j) \to \max;$$
$$\sum_j g_{sj}(x_j) = 0, \qquad s \in S,$$
$$x_j \in Q_j, \qquad j = 1, \ldots, J.$$

In this algorithm each iteration $\sigma - 1$ ends with an approximation of the primal and dual variables $x_j^{\sigma-1}$, $j = 1, \ldots, J$, $p_s^{\sigma-1}$, $s \in S$. The first stage of iteration σ consists of solving for each j the problem

$$\left\{ f_j(x_j) - \sum_{s \in S} p_s g_{sj}(x_j) - \frac{1}{2} \sum_{s \in S} q_s (g_{sj}(x_j) - g_{sj}(x_j^{\sigma-1}))^2 \right\} \to \max;$$
$$x_j \in Q_j,$$

the solution of which we denote by $\hat{x}_j^\sigma$. At the second stage we adjust the price vectors by the formula

$$\hat{p}_s^\sigma = p_s^{\sigma-1} + \alpha q_s \sum_{j=1}^{J} g_{sj}(\hat{x}_j^\sigma). \tag{2.4.17}$$

Then the final values for the primal and dual variables at iteration σ are computed from the formulas

$$x_j^\sigma = (1 - \beta) x_j^{\sigma-1} + \beta \hat{x}_j^\sigma, \qquad j = 1, \ldots, J;$$
$$p_s^\sigma = (1 - \beta) p_s^{\sigma-1} + \beta \hat{p}_s^\sigma, \qquad s \in S.$$

The authors of the algorithm prove that for $4/J > \alpha_0 > 0$, $0 < \beta_0 < \beta_1 < 1$, there exists a $q_0 > 0$ such that for some neighborhood W of the solution (x^*, p_0^*) of the original problem the algorithm will converge for any starting value $(x^0, p^0) \in W$ for any values of α, β, q_{sj} such that

$$q_{sj} > q_0, \qquad \alpha_0 < \alpha < \frac{4}{J}, \qquad \beta_0 < \beta < \beta_1.$$

The algorithm of Telle reduces solving (2.4.1)-(2.4.3) to finding the saddle point of the following modified Lagrangian function:

$$L(x,b,p) = \sum_j \left\{ f_j(x_j) - \sum_{s \in S} \frac{1}{2q_s} [(p_s + q_s(g_{sj} - b_{sj}))_{(+)}]^2 + \sum_{s \in S} \frac{1}{2q_s} p_s^2 \right\};$$

$$p_s \geqq 0, \qquad s \in S; \tag{2.4.18}$$

$$x_j \in Q_j, \qquad j = 1, \ldots, J; \tag{2.4.19}$$

$$\sum_j b_{sj} = b_s, \qquad s \in S, \tag{2.4.20}$$

where the function $L(x,b,p)$ is concave with respect to (x,b) and convex with respect to p. As before, the vector b is given by $b = (b_{sj} | s \in S, j = 1, \ldots, J)$.

Numerical experiments have shown that this algorithm had slightly worse convergence than the other algorithms. Telle in his algorithm introduces the function

$$M(b,p) = \max_{x \in Q} L(x,b,p), \tag{2.4.21}$$

where the set Q is formed as the direct product of the sets Q_j, $j = 1, \ldots, J$. The function $M(b,p)$ is convex with respect to b and concave with respect to p. Telle proves that the saddle point (b^*,p^*) determines the optimal resource allocation b_{sj}^* and the optimal dual prices p_s^*. Moreover, maximizing the function $L(x,b^*,p^*)$ over x in Q yields the triple (x^*,b^*,p^*), which is the saddle point of the function $L(x,b,p)$.

Telle gives a detailed analysis of the properties of the function $M(b,p)$. Since $M(b,p)$ is concave with respect to p and convex in b, we can use the generalized gradient method to calculate its saddle point (see Golstein, 1972).

The generalized gradient method for finding the saddle point of the differentiable function $M(b,p)$ of two vector variables b and p, with M concave in b and convex in p, where b runs through the set given by (2.4.20) and p given through $\mathbb{R}_+^{|S|}$, consists of generating the sequence $\{b^\sigma, p^\sigma\}$, according to

$$b^\sigma = Pr_{\mathbb{R}}(b^{\sigma-1} + \alpha^\sigma \nabla_b M(b^{\sigma-1}, p^{\sigma-1})), \tag{2.4.22}$$

$$p^\sigma = Pr_{\mathbb{R}_+^{|S|}}(p^{\sigma-1} - \beta^\sigma \nabla_p M(b^{\sigma-1}, p^{\sigma-1})). \tag{2.4.23}$$

Here $Pr_{\mathbb{R}}$ is the projection onto the set of feasible resource allocations $\mathbb{R}$, and $Pr_{\mathbb{R}_+^{|S|}}$ is the projection onto the set of feasible dual variables; $\nabla_b M(\beta p^{\sigma-1}, p^{\sigma-1})$ is the vector of partial derivatives of the function M with respect to b evaluated at the point $(b^{\sigma-1}, p^{\sigma-1})$; $\nabla_p M(b^{\sigma-1}, p^{\sigma-1})$ is the vector of partial derivatives of M with respect to p evaluated at the point $(b^{\sigma-1}, p^{\sigma-1})$; and the sequences α^σ, β^σ are such that α^σ, $\beta^\sigma > 0$ for all σ, $\lim \alpha^\sigma = 0$, and $\Sigma_\sigma \alpha^\sigma = \Sigma_\sigma \beta^\sigma = \infty$.

The main idea of Telle's algorithm is based on the fact that

$$\nabla_b M(b^{\sigma-1}, p^{\sigma-1}) = \nabla_b L(x^\sigma, b^{\sigma-1}, p^{\sigma-1}); \tag{2.4.24}$$

$$\nabla_p M(b^{\sigma-1}, p^{\sigma-1}) = \nabla_p L(x^\sigma, b^{\sigma-1}, p^{\sigma-1}), \tag{2.4.25}$$

where x^σ is the solution to the problem

$$L(x, b^{\sigma-1}, p^{\sigma-1}) \to \max; \tag{2.4.26}$$

$$x_j \in Q_j. \tag{2.4.27}$$

The equalities (2.4.24), (2.4.25) enable us to use the known function L(x,b,p) to find the saddle point of the function M(x,b), which is not known explicitly.

To complete the construction of the algorithm it is now sufficient to observe that

$$\frac{\partial L}{\partial b_{sj}}(x^\sigma, b^{\sigma-1}, p^{\sigma-1}) = (p_s^{\sigma-1} + q_s(g_{sj}(x_j^\sigma) - b_{sj}^{\sigma-1}))_{(+)}; \tag{2.4.28}$$

$$\frac{\partial L}{\partial p_s}(x^\sigma, b^{\sigma-1}, p^{\sigma-1}) = -\sum_j \left(\frac{p_s^{\sigma-1}}{q_s} + g_{sj}(x_j^\sigma) - b_{sj}^{\sigma-1}\right)_{(+)} + J\frac{p_s^{\sigma-1}}{q_s}. \tag{2.4.29}$$

Telle (1975) proved that for the convergence of the generalized gradient method to the saddle point of M(b,p) it is not always necessary that α^σ and β^σ tend to zero. He gave certain conditions (which unfortunately are cumbersome and nonconstructive, so that we do not present them here), and if these conditions are met, then for each problem there exist positive $\bar{\alpha}$, $\bar{\beta}$ such that if $\underline{\alpha} \le \alpha^\sigma \le \bar{\alpha}$ and $\underline{\beta} \le \beta^\sigma \le \bar{\beta}$ hold for all σ, the algorithm (2.4.24), (2.4.25) will converge to the saddle point of M(b,p). This method when usable converges much faster than does the algorithm with α^σ, β^σ tending to 0, as the latter methods generally are slow to converge. Furthermore, Telle proves that convergence is ensured if we impose the following conditions on α^σ, β^σ:

$$\alpha^\sigma < \left(\frac{2}{q_s}\ \frac{2q_s}{J}\right),$$

$$\beta^\sigma < \left(\frac{2}{q_s}\ \frac{2q_s}{J}\right).$$

Notice that the values of α, β here depend on s. If for some reason the values of the sequence α^σ, β^σ must be chosen on some other basis, the algorithm may diverge. Usually, α_s^σ is taken equal to $1/q_s$ and β_s^σ to q_s/J, but this choice does not always guarantee the convergence.

The transformations (2.4.22), (2.4.23), used in conjunction with (2.4.24), (2.4.25), (2.4.28), (2.4.29), yield the following formulas for Telle's algorithm when $\alpha_s^\sigma = 1/q_s$, $\beta_s^\sigma = q_s/J$:

$$p_s^\sigma = \frac{1}{J}\sum_{j=1}^{J}[p_s^{\sigma-1} + q_s(g_{sj}(x_j^\sigma) - b_{sj}^{\sigma-1})]_{(+)}; \tag{2.4.30}$$

$$b_{sj}^\sigma = \hat{b}_{sj}^\sigma + \frac{1}{J}\sum_{i=1}^{J}(b_s - \hat{b}_{sj}^\sigma); \tag{2.4.31}$$

$$\hat{b}_{sj}^{\sigma} = b_{sj}^{\sigma-1} + \left[\frac{p_s^{\sigma-1}}{q_s} + g_{sj}(x_j^{\sigma}) - b_{sj}^{\sigma-1}\right]_{(+)}. \tag{2.4.32}$$

Note that if resource s has positive dual variables in an optimal solution and the quantities p_s^{σ}, x_j^{σ} are located close to the optimum, the terms that take positive parts would all be positive and (2.4.30) and (2.4.32) would take the simpler forms:

$$p_s^{\sigma} = p_s^{\sigma-1} + \frac{q_s}{J}\left[\sum_{j=1}^{J} g_{sj}(x_j^{\sigma}) - b_s\right]; \tag{2.4.30'}$$

$$\hat{b}_{sj}^{\sigma} = g_{sj}(x_j^{\sigma}) + \frac{p_s^{\sigma-1}}{q_s}. \tag{2.4.32'}$$

Substitution of (2.4.32) into (2.4.31) yields

$$b_{sj}^{\sigma} = g_{sj}(x_j^{\sigma}) + \frac{1}{J}\sum_{j=1}^{J} [b_s - g_{sj}(x_j^{\sigma})]. \tag{2.4.31'}$$

Note that (2.4.30') coincides with (2.4.17) if we take $\alpha = 1/J$ in (2.4.17) and assume that $b_s = 0$. We can interpret (2.4.31) as an additional balancing of the quantities $g_{sj}(x_j^{\sigma})$ that represent the subsystems' needs for resources. These requirements are replaced by the quantities b_{sj}^{σ} according to (2.4.31) such that $\Sigma_j b_{sj}^{\sigma} = b_s$. Prilutsky (1979) has performed comparative numerical experiments with elastic allocation algorithms for linear programming problems. These experiments have shown that the algorithms based on modified Lagrangian functions are more efficient than the algorithms based on penalty functions. The algorithms of Telle and Razumikhin that explicitly consider the constraint (2.4.15) have proven to be the most effective of the algorithms based on the modified Lagrangian function.

PART II

ITERATIVE AGGREGATION ALGORITHMS FOR SOLVING SYSTEMS OF EQUATIONS

3

Iterative Aggregation Algorithms for Solving Systems of Interproduct Input-Output Model Equations

Although iterative aggregation algorithms have been developed for solving problems of large dimension with complex structure, it is easier to understand the principles of iterative aggregation by first analyzing models with a simple structure, of which input-output models are a particular example. The iterative aggregation methods for the solution of interproduct input-output models use some specific properties (see Section 3.1) of the models.

The first, so-called "basic" iterative aggregation methods (see Section 3.2-3.4), as well as their modifications (Section 3.6), manifested convergence in all numerical experiments and did so at a rate faster than the classical iterative processes for linear systems of equations (see Section 4.5). However, the proof of convergence for the "basic" iterative aggregation algorithms for solving input-output models has not been found in the general case. The proofs that are available (see Appendixes 3.1 and 3.2) are valid only for a number of special cases of varying practical importance. Similar results have also been obtained for various modifications of the basic processes (see Appendix 3.4).

At the same time some new iterative aggregation algorithms for solving interproduct input-output models have been proposed (see Section 3.5) that are guaranteed to converge globally (see Appendix 3.3).

Also, it is now known (see Section 3.7) how to transform the processes analyzed in Sections 3.2-3.4 and 3.6 into globally convergent algorithms.

3.1 INPUT-OUTPUT MODEL WITH VALUES OF PRODUCTS IN FIXED PRICES AND THE MODEL WITH VALUES OF PRODUCTS IN PHYSICAL UNITS

The following relations underlie both interproduct input-output model with fixed prices of products and input-output models with values of products in physical units:

$$x_g = \sum_{q=1}^{G} \hat{a}_{gq} x_q + \hat{b}_g, \qquad g = 1, \ldots, G; \tag{3.1.1}$$

$$\hat{a}_{gq} \geq 0, \quad g, q = 1, \ldots, G; \quad \hat{b}_g \geq 0, \quad g = 1, \ldots, G. \tag{3.1.2}$$

The subscripts g and q denote types of products (g, q = 1, ..., G), x_g the unknown output of the gth product, $\hat{b}_g$ the known final demand of the gth product (which includes personal and nonproductive consumption, fixed capital investment, increasing inventories, and foreign trade balance), and $\hat{a}_{gq}$ the known input coefficient of the gth product in producing the qth product.

The difference between physical and cost interproduct input-output models is not confined to the fact that in equation (3.1.1) in the first case the outputs x_g, the input coefficients $\hat{a}_{gq}$, and the quantities of final demand b_g are measured in physical units appropriate to the given product, whereas in the latter case output, coefficients, and final demand are measured in terms of cost. The matrix $(\hat{a}_{gq})$ of input coefficients has different properties depending on whether the units of measurement in the model are cost or physical units.

The cost of the products used for manufacturing a cost unit of a given product is naturally less than 1. Otherwise, the production would essentially be unprofitable since production costs also include labor wages and capital amortization. Therefore, for interproduct input-output models with fixed prices of products, these facts imply the following conditions on the input coefficients $\hat{a}_{gq}$:

$$\sum_g \hat{a}_{gq} < 1, \qquad q = 1, \ldots, G,$$

or equivalently,

$$\max_q \sum_g \hat{a}_{gq} = \alpha < 1. \qquad (3.1.3)$$

A similar condition applies to the interindustry input-output model as well.

For physical interproduct input-output models where the units of measurements are real physical quantities, the condition (3.1.3) generally does not hold. If, for example, it takes 2 tons of ore to produce 1 ton of steel, then letting g denote ore and q denote steel, we have $\Sigma_g \hat{a}_{gq} > 2$ in the specified units of measurement. Also, adding together coefficients expressed in different units of measurement produces quantities $\Sigma_g \hat{a}_{gq}$ that have no physical meaning.

However, there is a condition similar to but weaker then (3.1.3), which is valid for physical interproduct input-output models. Formally, this condition is: There exists a vector $p = (p_1,\ldots,p_G)$, $p_g \geq 0$, such that $\Sigma_g p_g \hat{a}_{gq} < p_q$ for all g = 1, ..., G. Or, equivalently, there exists a vector $p = (p_1,\ldots,p_g)$ such that

$$\max \sum_g p_g \hat{a}_{gq} p_q^{-1} = \alpha < 1. \qquad (3.1.3a)$$

Note that for p = (1,...,1) formulas (3.1.3a) and (3.1.3) are equivalent. Starting perhaps with Gale (1960), the condition (3.1.3a) for coefficient matrices of input-output models have come to be known as *productivity conditions*.

The condition that a coefficient matrix of a physical interproduct input-output model be productive has the economic meaning that there exist at least some prices at which the material costs for unit output of each type of product are less than the unit price of that product. This condition is a natural condition for physical interproduct input-output models. Also, it is equivalent to the following condition: The matrix A of nonnegative coefficients in (3.1.1) is productive if and only if there exists a positive vector x such that

$$x > Ax; \qquad (3.1.4)$$

that is, there exists a production plan in which more of each product is produced than is expended for production.

Another necessary and sufficient condition for the matrix A to be productive when (3.1.2) is valid is that for any nonnegative vector $\hat{b} = \{\hat{b}_g, g = 1, \ldots, G\}$ there exists a nonnegative solution to (3.1.1) or, equivalently, for any final demand vector there exists a production plan that can guarantee input-output equilibrium for this vector. This does not seem as natural a condition as the preceding two. However, it is this third condition that displays the importance of the concept of productivity in studying input-output equilibria, and we will be making use of it often in what follows.

There is one final condition that has a natural economic interpretation for deciding whether a matrix is productive. The matrix A is productive if and only if all $\hat{a}_{gq} \geq 0$ and the series

$$A + A^2 + A^3 + \cdots \tag{3.1.5}$$

converges. The sum of the series $A + A^2 + A^3 + \cdots$ is called the matrix of the total input coefficients. The condition (3.1.5) means that the matrix A is productive if and only if it is nonnegative and the total input coefficients matrix exists. This is equivalent to the condition that the matrix A is nonnegative and that the absolute values of the eigenvalues of A are less than 1. Various analogous definitions of productivity are examined in Gale (1960) and Vaksman (1969).

Computation of an interproduct input-output model is just a very simplified version of computing the plan for the entire economy. However, it is convenient to start our consideration of iterative aggregation procedures with this simple model.

3.2. BASIC TWO-LEVEL ITERATIVE AGGREGATION ALGORITHM FOR INTERPRODUCT INPUT-OUTPUT MODEL WITH VALUES OF PRODUCTS IN FIXED PRICES

We will first describe the two-level iterative aggregation process for interproduct input-output models (3.1.1)-(3.1.3) in the form in which it was originally proposed by Dudkin and Ershov (1965). One iteration of the process in this form consists of the following steps:

1. Fixing the present manufacturing intraindustry proportions, we compute the semiaggregate and aggregate input coefficients (the semiaggregate coefficient shows the input of a specific product for the manufacture of 1 unit of an industry output).
2. Using the aggregate coefficients, we solve the interindustry input-output model, which is of much lower dimension than the original problem.
3. Using the semiaggregate coefficients, we calculate the new approximation, which can be used to calculate new proportions in the semiaggregate and aggregate coefficients (i.e., to start the next iteration).

Formally, the algorithm is described as follows. The set $\{1, \ldots, G\}$ of indices denoting types of products is partitioned into J subsets G_i. Each subset G_i includes those types of products that are produced in the ith industry. We denote industries by the subscripts i, j (i, j = 1, ..., J). We assume that each type of product belongs to one and only one industry, that is, $\{1,\ldots,G\} = \bigcup_{i=1}^{J} G_i$, $G_i \cap G_j = \emptyset$. Let $x_g^{\sigma-1}$ denote the approximate solution to the original problem obtained after iteration $\sigma - 1$. One iteration of the algorithm proceeds as follows:

1. Compute the semiaggregate coefficients:

$$\tilde{a}_{gj}^{\sigma} = \frac{\sum_{q \in G_j} \hat{a}_{gq} x_q^{\sigma-1}}{\sum_{q \in G_j} x_q^{\sigma-1}}, \qquad g = 1, \ldots, G, \; j = 1, \ldots, J. \tag{3.2.1}$$

2. Compute the aggregate coefficients:

$$a_{ij}^{\sigma} = \sum_{g \in G_i} \tilde{a}_{gj}^{\sigma}, \qquad i, j = 1, \ldots, J. \tag{3.2.2}$$

3. Solve the interindustry input-output model:

$$X_i = \sum_j a_{ij}^{\sigma} X_j + b_i, \qquad b_i = \sum_{g \in G_i} \hat{b}_g, \qquad i = 1, \ldots, J, \tag{3.2.3}$$

and denote the solution by X_i^{σ}.

4. Compute the new approximation:

$$x_g^\sigma = \sum_j \tilde{a}_{gj}^\sigma X_j^\sigma + \hat{b}_g, \qquad g = 1, \ldots, G. \tag{3.2.4}$$

The values x_g^σ are then used to calculate the coefficients $\tilde{a}_{gj}^{\sigma+1}$ and $a_{ij}^{\sigma+1}$ to start the next iteration.

It is easy to show that this process is equivalent to the iterative aggregation process, described in the Introduction. Indeed, the semiaggregate coefficients (3.2.1) can be written as follows (h is an additional subscript denoting type of product):

$$\tilde{a}_{gj}^\sigma = \sum_{q \in G_j} \hat{a}_{gq} \frac{x_q^{\sigma-1}}{\Sigma_{h \in G_j} x_h^{\sigma-1}}, \qquad g = 1, \ldots, G, \; j = 1, \ldots, J. \tag{3.2.1a}$$

The fraction $x_q^{\sigma-1}/\Sigma_{h \in G_j} x_h^{\sigma-1}$ represents the specific weight of product q in the overall production of the industry j; it is this weight that is changed in the formulas from iteration to iteration. This allows us to describe the iterative aggregation algorithm without having to resort to the semiaggregate coefficients. Accordingly, having (3.2.1) in mind, the algorithm can be restated as follows:

1. Compute the aggregation weights:

$$d_q^\sigma = \frac{x_q^{\sigma-1}}{\Sigma_{h \in G_j} x_h^{\sigma-1}}, \qquad q \in G_j, \qquad j = 1, \ldots, J. \tag{3.2.5}$$

2. Compute the aggregate coefficients:

$$a_{ij}^\sigma = \sum_{g \in G_i} \sum_{q \in G_j} \hat{a}_{gq} d_q^\sigma, \qquad i, j = 1, \ldots, J. \tag{3.2.6}$$

3. Solve the interindustry input-output model:

$$X_i = \sum_j a_{ij}^\sigma X_j + b_i, \qquad b_i = \sum_{g \in G_i} \hat{b}_g, \qquad i = 1, \ldots, J. \tag{3.2.3}$$

and denote the solution by X_i^σ.

4. Compute the new approximation:

$$x_g^\sigma = \sum_j \sum_{q \in G_j} \hat{a}_{gq} d_q^\sigma X_j^\sigma + \hat{b}_g. \tag{3.2.7}$$

However, even this description can be simplified, as the algorithm can be described without the use of either the semiaggregate and aggregate coefficients or the specific product weights. This is achieved by forming an aggregate model with peculiar variables. Each of these variables z_i^σ shows the change of some industry output X_i^σ at iteration σ, in comparison with the volume of the same industry output $\Sigma_{g \in G_i} x_g^{\sigma-1}$ obtained at the previous iteration $\sigma - 1$, that is,

$$z_i^\sigma = \frac{X_i^\sigma}{\Sigma_{g \in G_i} x_g^{\sigma-1}}. \tag{3.2.8}$$

We can replace a_{ij}^σ on the right-hand side of the aggregate model (3.2.3) by the same quantity expressed in terms of the original model:

$$a_{ij}^\sigma = \sum_{g \in G_i} \tilde{a}_{gj}^\sigma = \frac{\Sigma_{g \in G_i} \Sigma_{q \in G_j} \hat{a}_{gq} x_q^{\sigma-1}}{\Sigma_{h \in G_j} x_h^{\sigma-1}}.$$

Then we can rewrite the interindustry input-output model (3.2.3) as follows:

$$X_i = \sum_j \sum_{q \in G_j} \hat{a}_{gq} x_q^{\sigma-1} \frac{X_i}{\Sigma_{h \in G_j} x_h^{\sigma-1}} + b_i.$$

After making these substitutions and recalling that

$$X_i = \frac{X_i}{\Sigma_{g \in G_i} x_g^{\sigma-1}} \sum_{g \in G_i} x_g^{\sigma-1},$$

we can write the aggregate model as

$$z_i \sum_{g \in G_i} x_g^{\sigma-1} = \sum_j \sum_{g \in G_i} \sum_{q \in G_j} \hat{a}_{gq} x_q^{\sigma-1} z_j + b_i.$$

It is easy to see that this system of equations is equivalent to the interindustry input-output model (3.2.3). Indeed, if z_i^σ is the solution of this system, then $X_i^\sigma = z_i^\sigma \Sigma_{g \in G_i} x_g^{\sigma-1}$ is the solution of the model (3.2.3); conversely, knowing the solution X_i^σ of (3.2.3) we can obtain z_i^σ from formula (3.2.8).

Let us also note that $X_i^\sigma = \Sigma_{g \in G_i} x_g^\sigma$. Indeed, if we sum up both the left and right sides of equations (3.2.4) concerning $g \in G_i$, we obtain

$$\sum_{g \in G_i} x_g^\sigma = \sum_{g \in G_i} \sum_j \tilde{a}_{gj}^\sigma X_j^\sigma + \sum_{g \in G_i} \hat{b}_g = \sum_j a_{ij}^\sigma X_j^\sigma + b_i = X_i^\sigma.$$

One iteration of the process stated in this form includes the following steps:

1. Compute the input-output flows and the industries' outputs from the solution at the previous iteration:

$$X_{ij}^{\sigma-1} = \sum_{g \in G_i} \sum_{q \in G_j} \hat{a}_{gq} x_q^{\sigma-1}, \qquad X_i^{\sigma-1} = \sum_{g \in G_i} x_g^{\sigma-1}. \tag{3.2.9}$$

2. Solve the system of interindustry input-output equations:

$$X_i^{\sigma-1} z_i = \sum_j X_{ij}^{\sigma-1} z_j + b_i, \qquad b_i = \sum_{g \in G_i} \hat{b}_g, \tag{3.2.10}$$

   and denote the solution of (3.2.10) by z_i^σ.

3. Compute the new approximation:

$$x_g^\sigma = \sum_j z_j^\sigma \sum_{q \in G_j} \hat{a}_{gq} x_q^{\sigma-1} + \hat{b}_g. \tag{3.2.11}$$

The two-level iterative aggregation algorithm given by (3.2.9)-(3.2.11) differs from the algorithm described by (3.2.1)-(3.2.4) in that in this case the proportions between the outputs of specific products are fixed directly in each input flow $X_{ij}^{\sigma-1}$ of one industry's products for the production of another industry's products and in each industry output $X_i^{\sigma-1}$.

It can be seen that equations (3.2.9) and (3.2.11) can be implemented at the level of the individual industries, while (3.2.10) must be implemented at the level of the central planning body. Therefore, at each iteration only the aggregate indices are passed from the industries to the central planning body, and vice versa. This feature of only passing aggregate information between levels is shared by all other two-level and multilevel iterative aggregation algorithms.

The most straightforward interpretation of how information is being passed in (3.2.9)-(3.2.11) is as follows. Injustry j computes (3.2.9) and passes to the center the quantities X_{ij}, i, j = 1, ..., J, and X_i, i = 1, ..., J. The center receives $J^2 + J$ such numbers from the industries. The center then computes (3.2.10) and passes to industry j the quantity z_j^{σ}. Industry j then uses this to compute the demand $x_{gj}^{\sigma-1} = \Sigma_{q \in G_j} \hat{a}_{gq} z_j^{\sigma} x_q^{\sigma-1}$, j = 1, ..., J, for each product and passes the quantities $x_{gj}^{\sigma-1}$ to industry i, which solves (3.2.11). Thus the center computes (3.2.10) and passes on J numbers to the industries and then the industries pass on (J - 1)G numbers $x_{gj}^{\sigma-1}$ in the process of computing (3.2.11).

There exist convergence proofs for the process (3.2.9)-(3.2.11) for various special cases (see Appendix 3.1), but there is no proof of convergence for the general case. However, one of these cases—Shchennikov's proof—allows us to use the algorithms in practice with confidence in its convergence. Shchennikov (1965) has proven convergence for the iterative aggregation process that aggregates the original system into one equation with one unknown at each iteration. To use the algorithm with confidence in its convergence, it appears at first that we must concentrate all the initial information $\hat{a}_{gq}$, g, q = 1, ..., G, $\hat{b}_g$, g = 1, ..., G, in one place (the center). However, it is not so. Indeed, at each iteration σ each subsystem (industry) j can pass on to the center the aggregate outputs $X_j^{\sigma-1}$ and flows $X_{ij}^{\sigma-1}$, whereas the center can aggregate the system of equations (3.2.10) into a single equation with one unknown, z^σ. The center can then solve this equation, disaggregate this solution down to the industry level, and pass to the industries the quantities X_j^σ, which the industries disaggregate into the quantities x_g^σ. Thus even though a mathematical proof has not been found for the general case, this does not limit the possibilities of using the iterative aggregation algorithm in hierarchical systems.

3.3 BASIC ONE-LEVEL ITERATIVE AGGREGATION ALGORITHM FOR INTERPRODUCT INPUT-OUTPUT MODEL WITH VALUES OF PRODUCTS IN FIXED PRICES

The basic one-level iterative aggregation algorithm for solving (3.1.1)-(3.1.3) was proposed by Vakhutinsky et al. (1975). In this algorithm at each iteration we construct J problems, each corresponding to an individual industry j and do not construct a separate aggregated model. However, each industry model has two groups of variables: (1) variables that represent outputs of specific products of that industry to which the given model corresponds (this group of variables coincides with variables in the original model), and (2) variables that represent aggregate outputs of other industries. In other words, in problem j all products except for those of industry j are aggregated.

Each industry model includes two types of interconnected equations: interproduct input-output equations concerning all specific products of this industry, and interindustry input-output equations concerning all the industries.

The process is initialized by calculating the semiaggregate coefficients of input of industries production for the manufacture of a unit output of a specific product:

$$\tilde{\tilde{a}}_{iq} = \sum_{g \in G_i} \hat{a}_{gq}, \qquad i = 1, \ldots, j-1, j+1, \ldots, J, \quad q = 1, \ldots, G, \tag{3.3.1}$$

as well as the aggregate indices of final demand for the products of entire industries:

$$b_i = \sum_{g \in G_i} \hat{b}_g.$$

At the start of iteration σ we are given the vectors $x_g^{\sigma-1}$, $g = 1, \ldots, G$, as the initial approximation to the outputs of products. Then one iteration of the algorithm proceeds as follows:

1. Compute the semiaggregate coefficients of each specific product input for the manufacture of a unit output of each industry:

$$\tilde{a}_{gj}^{\sigma} = \sum_{q\in G_j} \hat{a}_{gq} \frac{x_q^{\sigma-1}}{\Sigma_{h\in G_j} x_h^{\sigma-1}}, \qquad g = 1, \ldots, G, \qquad j = 1, \ldots, J. \tag{3.3.2}$$

2. Compute the aggregate input coefficients:

$$a_{ij}^{\sigma} = \sum_{g\in G_i} \tilde{a}_{gj}^{\sigma}, \qquad i, j = 1, \ldots, J. \tag{3.3.3}$$

3. Solve for each industry j the local problem (τ is also a subscript denoting some industry):

$$\begin{aligned} &\sum_{q\in G_j} \hat{a}_{gq} x_q + \sum_{\tau\neq j} \tilde{a}_{g\tau} X_\tau^j + \hat{b}_g = x_g, && g \in G_j; \\ &\sum_{q\in G_j} \tilde{\tilde{a}}_{iq} x_q + \sum_{\tau\neq j} a_{i\tau}^{\sigma} X_\tau^j + b_i = X_i^j, && i = 1, \ldots, j-1, \\ & && j+1, \ldots, J, \end{aligned} \tag{3.3.4}$$

and denote the solution of (3.3.4) by $X_\tau^{j,\sigma}$, x_q^{σ}, $\tau = 1, \ldots, j-1, j+1, \ldots, J$, $q \in G_j$, where $X_\tau^{j,\sigma}$, $\tau \neq j$, denotes the output of industry τ to be determined in industry j.

The quantities x_q^{σ}, $q \in G_j$, derived by calculating one iteration of the algorithm are the σth approximation of the output from industry and serve as the initial values for the next iteration. Since the index j ranges from 1 to J, the approximations for the variables of the original problems will be found when calculating iteration σ.

This algorithm can also be described in terms of flows in some system as follows:

1. Compute the semiaggregate and aggregate input-output flows and aggregate outputs:

$$x_{gj}^{\sigma-1} = \sum_{q \in G_j} \hat{a}_{gq} x_q^{\sigma-1}; \quad X_{ij}^{\sigma-1} = \sum_{g \in G_i} x_{gj}^{\sigma-1}; \quad X_i^{\sigma-1} = \sum_{g \in G_i} x_g^{\sigma-1}. \tag{3.3.5}$$

2. For each industry j solve the local problem:

$$x_g = \sum_{q \in G_j} \hat{a}_{gq} x_q + \sum_{\tau \neq j} x_{g\tau}^{\sigma-1} z_\tau^j + \hat{b}_g, \qquad g \in G_j,$$
$$X_i^{\sigma-1} z_i^j = \sum_{q \in G_j} \tilde{\tilde{a}}_{iq} x_q + \sum_{\tau \neq j} X_{i\tau}^{\sigma-1} z_\tau^j + b_i, \qquad i = 1, \ldots, j-1, \; j+1, \ldots, J. \tag{3.3.6}$$

Here the variables z_i^j tell us by how much each industry must change its level of output given the fixed output structure of that industry in order for j's industry model to be in equilibrium. The values of the detailed variables x_g^σ obtained from the local models are then used to initiate the next iteration, and are discarded.

Although in the one-level iterative aggregation algorithms local problems involve both aggregate and nonaggregate terms, it is still true that the interindustry input-output proportions are solved in each subproblem with the intraindustry proportions fixed except for the particular industry under consideration.

3.4 ITERATIVE AGGREGATION ALGORITHMS FOR THE SOLUTION OF INTERPRODUCT INPUT-OUTPUT MODEL WITH VALUES OF PRODUCTS IN PHYSICAL UNITS

If we try to use the algorithm (3.2.1)-(3.2.4) for solving an arbitrary interproduct input-output model with products measured in physical units, we are confronted with an unrealistic situation, that is, the necessity for adding together tons and kilometers, for example. In this case the algorithm may be divergent and the aggregate problem may not be solvable (see Dudkin and Rabinovich, 1976). However, a close inspection of (3.2.1)-(3.2.4) reveals how economically meaningful operations can be performed for this problem. This is done by involving such fixed prices p_g, $g = 1, \ldots, G$, in the process of solving (3.1.1), (3.1.2) that the property of productivity (3.1.3a) is satisfied with these prices.

Let $\hat{x}^{\sigma-1}$ be the approximate solution to the original problem at the end of iteration $\sigma - 1$. Then one iteration of the process of solving (3.1.1), (3.1.2), (3.1.3a) proceeds as follows:

1. Compute the semiaggregate coefficients:

$$\tilde{a}^{\sigma}_{gj} = \frac{\sum_{q \in G_j} \hat{a}_{gq} \hat{x}^{\sigma-1}_q}{\sum_{q \in G_j} p_q \hat{x}^{\sigma-1}_q}. \tag{3.4.1}$$

2. Compute the aggregate coefficients:

$$a^{\sigma}_{ij} = \sum_{g \in G_i} p_g \tilde{a}^{\sigma}_{gj}. \tag{3.4.2}$$

3. Solve the aggregate input-output problem:

$$X_i = \sum_j a^{\sigma}_{ij} X_j + b_i, \qquad b_i = \sum_{g \in G_i} p_g \hat{b}_g, \tag{3.4.3}$$

and denote the solution of (3.4.3) by X^{σ}_i.

4. Compute the new approximation of variables:

$$\hat{x}^{\sigma}_g = \sum_j \tilde{a}^{\sigma}_{gj} X^{\sigma}_j + \hat{b}_g. \tag{3.4.4}$$

The algorithm has a simpler structure when described in terms of flows:

1. Compute the semiaggregate flows:

$$\hat{x}^{\sigma-1}_{gj} = \sum_{q \in G_j} a_{gq} \hat{x}^{\sigma-1}_q. \tag{3.4.5}$$

2. Compute the aggregate flows and plans:

$$X^{\sigma-1}_{ij} = \sum_{g \in G_i} p_g \hat{x}^{\sigma-1}_{gj}; \tag{3.4.6}$$

$$X^{\sigma-1}_i = \sum_{g \in G_i} p_g \hat{x}^{\sigma-1}_g. \tag{3.4.7}$$

3. Solve the interindustry input-output model:

$$X_i^{\sigma-1} z_i = \sum_j X_{ij}^{\sigma-1} z_j + b_i \tag{3.4.8}$$

and denote the solution by z_i^σ.

4. Compute the new approximation of variables:

$$\hat{x}_g^\sigma = \sum_j x_{gj}^\sigma z_j^\sigma + \hat{b}_g. \tag{3.4.9}$$

It can be seen that this algorithm differs from the flow algorithm described in Section 3.2 only by the use of a somewhat more complex formula for $X_{ij}^{\sigma-1}$ and $X_i^{\sigma-1}$ at each iteration. Also, this algorithm is equivalent to using (3.2.1)-(3.2.4) to solve the interproduct input-output model (3.1.1)-(3.1.3), where instead of coefficients $\hat{a}_{gq}$ we use the coefficients $p_g a_{gq} p_q^{-1}$ and the quantities $p_g \hat{b}_g$ instead of $\hat{b}_g$:

$$x_g = \sum_q (p_g \hat{a}_{gq} p_q^{-1}) x_q + p_g \hat{b}_g, \qquad g = 1, \ldots, G, \tag{3.4.10}$$

where

$$\sum_q p_g \hat{a}_{gq} p_q^{-1} < 1, \qquad g = 1, \ldots, G.$$

By "equivalence" we mean that at each iteration of the two-level algorithm (3.4.5)-(3.4.9) for the interproduct input-output model with physical units of products [i.e., with an arbitrary productive matrix (a_{gq})] the quantities z_i^σ are identical to the same quantities calculated for the interproduct input-output model (3.4.10) with the changed productivity matrix $p_g \hat{a}_{gq} p_q^{-1}$. The quantities x_g^σ of this model (3.4.10) are related to the variables $\hat{x}_g^\sigma$ of the model with physical units of products by the formula $x_g^\sigma = p_g \hat{x}_g^\sigma$, $g = 1, \ldots, G$, provided that the two processes started with the same initial condition, that is, $x_g^0 = p_g \hat{x}_g^0$, $g = 1, \ldots, G$.

Therefore, the proof of convergence of the algorithm (3.2.5), (3.2.6), (3.2.3), (3.2.7) for the problem (3.1.1)-(3.1.3) given in Appendix 3.1 also works for the algorithm (3.4.5)-(3.4.9) for the problem (3.1.1)-(3.1.3a).

In comparison with the algorithm (3.2.9)-(3.2.11), the algorithm described by (3.4.5)-(3.4.9) requires G(J + 1) additional multiplications and (G - J)(J + 1) additional additions at each iteration (the values of b_i can be computed in advance). If, instead, we first transform the input-output model in physical units into a model with fixed prices and solve this using (3.2.9)-(3.2.11), the solution of the problem requires G^2 multiplications and G^2 divisions only once before the first iteration. Therefore, it is preferable to use (3.4.5)-(3.4.9) as long as the algorithm converges at a fairly fast rate and if J is small in comparison with G. Also, the direct application of (3.4.5)-(3.4.9) is often convenient in terms of the economic realities of the system.

Similarly, the one-level process (3.3.1)-(3.3.4) for the model (3.1.1), (3.1.2), (3.1.3a) consists of the following steps:

1. Compute the semiaggregate input-output flows and aggregate outputs:

$$\hat{x}_{gj}^{\sigma-1} = \sum_{q \in G_j} \hat{a}_{gq} \hat{x}_q^{\sigma-1}, \qquad X_{ij}^{\sigma-1} = \sum_{g \in G_i} p_g \hat{x}_{gj}^{\sigma-1},$$
$$X_i^{\sigma-1} = \sum_{g \in G_i} p_g x_g^{\sigma-1}. \tag{3.4.11}$$

2. Solve for each industry j the local combined interproduct and interindustry model:

$$\hat{x}_g = \sum_{q \in G_j} \hat{a}_{gq} \hat{x}_q + \sum_{\tau \neq j} \hat{x}_{g\tau}^{\sigma-1} z_\tau^j + \hat{b}_g, \qquad g \in G_j;$$
$$x_i^{\sigma-1} z_i^j = \sum_{q \in G_j} \tilde{a}_{iq} \hat{x}_q + \sum_{\tau \neq j} X_{i\tau}^{\sigma-1} z_\tau^j = b_i, \tag{3.4.12}$$
$$i = 1, \ldots, j - 1, j + 1, \ldots, J.$$

where

$$\tilde{a}_{iq} = \sum_{g \in G_i} p_g \hat{a}_{gq}, \qquad b_i = \sum_{g \in G_i} p_g \hat{b}_g.$$

An interesting algorithm that uses aggregation-disaggregation transformation only at one step of the successive approximation process was proposed by Manove and Weitzman (1978). They start with an initial approximation x^0 to the solution x^* of the equation $Ax + b = x$. Then they perform just one specific aggregation-disaggregation step, similar to the aggregation-disaggregation step in our algorithm (3.4.5)-(3.4.9) described above and obtain a new approximation, x^1.

It should be noted that they aggregate to one sector ($J = 1$). So the new approximation x^1 is proportional to the initial approximation x^0 (i.e., $x^1 = zx^0$). Then they proceed by successively applying simple iterations: $x^{\sigma+1} = Ax^\sigma + b$.

The main two differences between their single aggregation-disaggregation step and our aggregation-disaggregation procedure at every step of the process (3.4.5)-(3.4.8) are the following.

1. Instead of solving the aggregated problem

   $$(p, zx^0) = (p, Azx^0) + (p, b)$$

   they solve another problem:

   $$(p, zx^0) = (p, azx^0) + (p, b),$$

   where a is the aggregate economy-wide input-output coefficient. These two equations in z coincide if $A'p = ap$, that is, if the vector of prices p is an eigenvector of the matrix A' and a is the corresponding eigenvalue (matrices A and A' have the same eigenvalues).
2. Instead of using the vector p of current (or arbitrary productive) prices, they propose to find the "optimal" vector of prices ensuring the best approximation to the situation of $x = Ax + b$ at their single aggregation-disaggregation step. Manove and Weitzman prove that the eigenvector p of matrix A', corresponding to the maximal eigenvalue, gives such an approximation. Such a vector of prices was used previously by von Neumann in his model of expanding economy.

   The exact values of p and a are difficult to find since it would involve solving the characteristic equation $\det(A - kI) = 0$, where I is the identity matrix. Therefore, it is reasonable to approximate them by performing several iterations (one or two, for example) of the following procedure:

a. Choose an arbitrary strictly positive vector c.

b. Starting with the vector p^0 of current (or arbitrary productive) prices, successively define new vectors of prices by the formulas

$$p^\sigma = \frac{(p^{\sigma-1},c)}{(p^{\sigma-1},Ac)} (A'p^{\sigma-1}). \qquad (3.4.13)$$

Let us explain the meaning of this iterative process of approximating the vector of the von Neumann prices. Let $k_1, \ldots, k_n$ be the eigenvalues of A' and $p_1, \ldots, p_n$ be the corresponding eigenvectors, that is, $A'p_j = k_j p_j$, where $|k_1| = \max |k_j|$. We would like to approximate the vector p_1 of the von Neumann prices. Start with the vector p of current (or arbitrary productive) prices. It can be represented as a linear combination of p_j:

$$p = \Sigma_j \lambda_j p_j.$$

Then

$$pA^\sigma = \sum_j \lambda_j k_j^\sigma p_j = k_1^\sigma \sum_j \lambda_j \left(\frac{k_j}{k_1}\right) p_j.$$

Since $|k_j| < |k_1|$ for $j \neq 1$, we have

$$\lim_{\sigma\to\infty} \frac{(A')^\sigma p}{k_1^\sigma} = \lambda_1 p_1. \qquad (3.4.14)$$

So the direction of $(A')^\sigma p$ converges to the direction of p_1. We can see from (3.4.14) that the norm (length) of $(A')^\sigma p$ may converge to 0 (if $k_1 < 1$) or to ∞ (if $k_1 > 1$). However, if we replace $(A')^\sigma p$ by its multiple $\mu_\sigma (A')^\sigma p$ satisfying the condition $(\mu_\sigma (A')^\sigma p,c) = (p,c)$ (where c is an arbitrary positive vector), then $(A')^\sigma p$ converges to a multiple of p_1. The vector p^σ defined in (3.4.13) is precisely the vector $\mu_\sigma (A')^\sigma p$. Thus it converges to a multiple of p_1.

We can use the approximation of von Neumann prices not only for providing a single aggregation-disaggregation step as Manove and Weitzman proposed, but also at every step of iterative aggregation process (3.4.5)-(3.4.9).

We can combine our iterative aggregation process (3.4.5)-(3.4.9) with the Manove-Weitzman approximation of von Neumann prices in one of the following ways. The first algorithm starts with performing several iterations of the process (3.4.13). It yields an approximation p^σ to the vector of von Neumann prices. Then we continue with the process (3.4.5)-(3.4.9) using the price vector p^σ instead of the vector p of current prices at each iteration of the process.

In general, it is not true that the vector of von Neumann prices provides the fastest convergence. A counterexample is given in Appendix 3.2. However, for the vector of von Neumann prices, we can obtain a simple estimate of the rate of convergence. Namely, the rate of convergence of the algorithm (3.4.5)-(3.4.9) with the vector of von Neumann prices is determined by the second-largest eigenvalue of the matrix A, and this is better than the rate of convergence of the method of simple iterations. The latter is determined by the largest eigenvalue of the matrix A.

The second algorithm combines the iterative aggregation process (3.4.5)-(3.4.9) and the process of approximating the vector of von Neumann prices. It proceeds as follows. Start with an initial approximation x^0 and an initial vector of current or arbitrary prices p^0. Let $x^{\sigma-1}$ and $p^{\sigma-1}$ be approximations to the solution x* of the original problem and to the vector p* of von Neumann prices. Then one iteration of the process proceeds as follows:

1. Compute the aggregate volumes:

$$X_i^{\sigma-1} = \sum_{g \in G_i} p_g^{\sigma-1} x_g^{\sigma-1}, \qquad X_{ij}^{\sigma-1} = \sum_{g \in G_i} \sum_{q \in G_j} p_g^{\sigma-1} \hat{a}_{gq} x_q^{\sigma-1},$$

$$b_i^{\sigma-1} = \sum_{g \in G_i} p_g^{\sigma-1} \hat{b}_g. \tag{3.4.15}$$

2. Solve the aggregate problem:

$$z_i X_i^{\sigma-1} = \sum_{j=1}^{J} X_{ij}^{\sigma-1} z_j + b_i^{\sigma-1} \tag{3.4.16}$$

and denote the solution by z_j^σ.

3. Compute the new approximation of variables:

$$x_g^\sigma = z_i^\sigma \sum_g \hat{a}_{gq} x_q^{\sigma-1} + \hat{b}_g. \tag{3.4.17}$$

4. Compute the new vector of prices:

$$p_q^\sigma = \frac{\Sigma_g\, p_g^{\sigma-1} \hat{a}_{gq}}{\Sigma_g\, p_g^{\sigma-1}}. \tag{3.4.18}$$

In Appendix 3.2 we prove the convergence of the iterative aggregation algorithm to the solution of interproduct input-output model (3.1.1), (3.1.2), (3.1.3a) with prices tending to a limit for the case $J = 1$. Obviously, this proves the convergence of the algorithm (3.4.15)-(3.4.18) when $J = 1$.

3.5 TWO-LEVEL ITERATIVE AGGREGATION PROCESSES WITH THE MINIMIZATIONS OF IMBALANCES IN A SEMIAGGREGATE PROBLEM

There exist different methods for determining the output of each industry when intraindustry proportions are fixed. In the algorithms discussed previously, at each iteration the industry outputs are determined by requiring that they are in equilibirum; that is, the output of each industry is utilized (under implicit conditional assumption that the intraindustry proportions have been set correctly).

However, when solving the interindustry input-output model the intraindustry proportions are actually fixed only approximately. Having this in mind, we may desire at each iteration to find the output by industry that would produce the best possible balance of outputs at the lowest level while assuming fixed intraindustry proportions. This idea leads us to a new iterative aggregation process.

This new algorithm is applied to solving an interproduct input-output model with weaker conditions for initial information. Specifically, the algorithm described below fits the general interproduct input-output model (3.1.1):

$$x_g = \sum_{q=1}^{G} \hat{a}_{gq} x_q + \hat{b}_g, \qquad g = 1, \ldots, G,$$

where the nonnegativity constraint (3.1.2) is no longer required to hold. The convergence of this algorithm is proved under certain restrictions on the matrix $(\hat{a}_{gq})$, which are weaker than the productivity conditions (3.1.3a).

Negative coefficients $\hat{a}_{gq}$, $g \neq q$, and quantities $\hat{b}_g$, $g = 1, \ldots,$ G, can have economic sense. Indeed, since imports and inventories of certain commodities can exceed positive final demand for these commodities, the vector of final demand may have negative components. Also, there are often by-products associated with the production of certain commodities and therefore the coefficients $\hat{a}_{gq}$, $g \neq q$, can be negative (a negative coefficient shows the output of product g as a by-product of the production of one unit of the product q).

In the algorithm (3.2.9)-(3.2.11) the interindustry input-output model is actually computed at each iteration as an intermediate step of the algorithm when determining a new approximation for the variables of the original problem (3.1.1). This intermediate step coordinates the balance between specific products $g = 1, \ldots, G$ realized at each iteration. However, after computing the semiaggregate flows $x_{gj}^{\sigma-1} = \Sigma_{q \in G_j} \hat{a}_{gq} x_q^{\sigma-1}$ for iteration σ, we could replace (3.2.10) by the following system of input-output equations with semiaggregate flows:

$$x_g^{\sigma-1} z_i - \sum_{j=1}^{J} x_{gj}^{\sigma-1} z_j = \hat{b}_g, \qquad g \in G_i, \quad i = 1, \ldots, J. \qquad (3.5.1)$$

This is a system of G linear equations in J variables with G > J. Hence the system is overdetermined and in general does not have a solution that satisfies all the equations.

If the proportions between the outputs of specific products, fixed for each industry, are not the correct ones, we could not find the outputs of industries that would ensure observance of interproduct balances. However, as soon as the detailed product proportions are fixed, it is reasonable to try and find the aggregate outputs that would lessen the imbalances in some sense. As noted in Section 1.1, the imbalance (i.e., the failure to satisfy the system of equations) can be characterized by the norm of the imbalances vector.

The idea behind the present algorithm can be described as follows: With fixed intraindustry proportions, we decrease the norm of the imbalances vector by changing the production volumes at the industry level; that is, we find numbers z_i, $i = 1, \ldots, J$, such that the plan $\tilde{x}^\sigma = \{x_g^{\sigma-1} z_i,\ g \in G_i,\ i = 1, \ldots, J\}$ is more balanced than the plan $x^{\sigma-1}$; then we fix the production volumes at the industry level and solve for the output volumes at the product level.

A detailed description of the algorithm to find the z_i so as to minimize the Euclidean norm of the imbalances vector is as follows. One iteration of the algorithm will consist of the following steps:

1. Compute the semiaggregate input flows of the products for the industry outputs:

$$x_{gj}^{\sigma-1} = \sum_{q \in G_j} \hat{a}_{gq} x_q^{\sigma-1}. \tag{3.5.2}$$

2. Solve the problem of minimization of imbalances in the model (3.5.1):

$$\min \Delta^2 = \min \sum_i \sum_{g \in G_i} \left[z_i (x_{gi}^{\sigma-1} - x_g^{\sigma-1}) + \sum_{j \neq i} z_j x_{gj}^{\sigma-1} + \hat{b}_g \right]^2,$$

or more simply:

$$\min \Delta^2 = \min \sum_i \sum_{g \in G_i} \left(\sum_j z_j x_{gj}^{\sigma-1} - z_i x_g^{\sigma-1} + \hat{b}_g \right)^2 \tag{3.5.3}$$

and denote the solution of (3.5.3) by z_i^σ.

3. Compute the new approximation by equation (3.2.11):

$$x_g^\sigma = \sum_j \sum_{q \in G_j} \hat{a}_{gq} z_j^\sigma x_q^{\sigma-1} + \hat{b}_g. \tag{3.5.4}$$

Let $D^{\sigma-1}$ be the matrix of semiaggregate input-output flows with elements

$$d_{gj}^{\sigma-1} = \begin{cases} x_{gj}^{\sigma-1}, & g \notin G_j \\ x_{gj}^{\sigma-1} - x_g^{\sigma-1}, & g \in G_j \end{cases} \qquad (3.5.5)$$

Then equation (3.5.3) can be rewritten as

$$\min \Delta^2 = \min \sum_g \left(\sum_j d_{gj}^{\sigma-1} z_j + \hat{b}_g\right)^2 .$$

The minimum of any quadratic function is reached at the point in which all the partial derivatives of this function are equal to zero. Hence solving (3.5.3) is equivalent to solving the system of equations

$$\frac{\partial(\Delta^2)}{\partial z_i} = \frac{\partial(\Sigma_g(\Sigma_j\ d_{gj}^{\sigma-1} z_j + \hat{b}_g)^2)}{\partial z_i} = 0.$$

After partial differentiation we obtain the system

$$\Sigma_g\ d_{gi}^{\sigma-1}(\Sigma_j\ z_j d_{gj}^{\sigma-1} + \hat{b}_g) = 0, \qquad i = 1, \ldots, J.$$

It has been proven that if the Euclidean norm of the matrix A in this algorithm is less than 1, the imbalance $(Ax^\sigma + b - x^\sigma)$ monotonically approaches zero at a rate not slower than a geometric progression with denominator $\|A\|$. Note that this process can be applied to arbitrary matrix A and vector b.

The vector z^σ is obtained from solving the system

$$(D^{\sigma-1})'D^{\sigma-1}z = -(D^{\sigma-1})'b. \qquad (3.5.6)$$

Let us compare the algorithm (3.5.2), (3.5.5), (3.5.6), (3.5.4) with the preceding ones. The system (3.2.10) can be rewritten as

$$TD^{\sigma-1}z = -Tb, \qquad (3.5.7)$$

where $T = (t_{iq})$, $i = 1, \ldots, J$, $q = 1, \ldots, G$, $t_{iq} = 1$ if $q \in G_i$ and $t_{iq} = 0$ otherwise. Hence (3.2.10) differs from (3.5.6) only in that (3.2.10) is derived by aggregating the rows of (3.5.2) using the matrix T, whereas in (3.5.6) we use the matrix $(D^{\sigma-1})'$.

It is easy to see that in (3.5.6) we solve an input-output model with semiaggregate flows D^{σ}, while in (3.2.10) we solve a model with aggregate flows and outputs.

This may give the impression that the algorithm (3.5.2), (3.5.5), (3.5.6), (3.5.4) requires passing the G•J matrix of semiaggregate flows between levels at each iteration. However, the computation of (3.5.6) can be broken down into separate stages, some of which can be implemented at the lower levels of control, while the upper level of control deals only with a J•J matrix of aggregate flows. The matrix $D^{\sigma-1}$ has a natural partition into submatrices $D_{\tau}^{\sigma-1}$, $\tau = 1, \ldots, J$, where the submatrix $D_{\tau}^{\sigma-1} = (d_{gi}^{\sigma-1})$, $g \in G_{\tau}$; that is, it contains all rows of the matrix $D^{\sigma-1}$ that have indexes $g \in G_{\tau}$. With this notation we can see that

$$(D^{\sigma-1})'D^{\sigma-1} = \sum_{\tau=1}^{J} (D_{\tau}^{\sigma-1})'D_{\tau}^{\sigma-1} \tag{3.5.8}$$

and

$$(D^{\sigma-1})'b = \sum_{\tau=1}^{J} (D_{\tau}^{\sigma-1})'b_{\tau}^{\sigma}, \tag{3.5.9}$$

where b_{τ}^{σ} is a subvector of the vector b containing the components whose indexes are in G_{τ}.

Let $H_{\tau}^{\sigma-1} = (h_{ij\tau}^{\sigma-1})$ denote $(D_{\tau}^{\sigma-1})'D_{\tau}^{\sigma-1}$ and $b_{\tau}^{\sigma-1} = (b_{i\tau}^{\sigma-1})$ denote $(D_{\tau}^{\sigma-1})'b_{\tau}$. Then it is clear that

$$h_{ij\tau}^{\sigma-1} = \sum_{g \in G_{\tau}} d_{gi}^{\sigma-1} d_{gj}^{\sigma-1}, \tag{3.5.10}$$

$$b_{i\tau}^{\sigma-1} = \sum_{g \in G} d_{gi}^{\sigma-1} \hat{b}_g. \tag{3.5.11}$$

Note that since the $H_{\tau}^{\sigma-1}$ are symmetric matrices $(h_{ij\tau}^{\sigma-1} = h_{ji\tau}^{\sigma-1})$, it is sufficient to compute $h_{ij\tau}^{\sigma-1}$ for $j \geq i$; that is, it is sufficient to compute the upper right triangular submatrix of $H_{\tau}^{\sigma-1}$, including diagonal elements. Therefore, each subsystem τ computes and passes to the center the $J(J + 1)/2$ parameters $h^{\sigma-1}_{ij\tau}$, $j \geq i$, as well as the J parameters $b^{\sigma-1}_{i\tau}$.

The center then proceeds by computing the matrix $(D^{\sigma-1})'D^{\sigma-1}$ with elements $h^{\sigma-1}_{ij}$ and the vector $(D^{\sigma-1})'b$ with components $b^{\sigma-1}_{\tau}$ from the formulas (3.5.8), (3.5.9) using the matrices $(D^{\sigma-1}_{\tau})'D^{\sigma-1}_{\tau}$ and vectors $(D^{\sigma-1}_{\tau})'b$ furnished by the subsystems

$$h^{\sigma-1}_{ij} = \sum_{\tau=1}^{J} h^{\sigma-1}_{ij\tau}, \tag{3.5.12}$$

$$b^{\sigma-1}_{i} = \sum_{\tau=1}^{J} b^{\sigma-1}_{i\tau}. \tag{3.5.13}$$

Then the center solves the system of equations

$$\sum_{j=1}^{J} h^{\sigma-1}_{ij} z_j = b^{\sigma-1}_{i}, \qquad i = 1, \ldots, J. \tag{3.5.14}$$

Thus the process (3.5.2), (3.5.5), (3.5.6), (3.5.4) is transformed into the process (3.5.2), (3.5.5), (3.5.10)-(3.5.14), (3.5.4).

The upper level of the hierarchy performs only the computations (3.5.12)-(3.5.14) involving aggregate indices, while the lower-level management boards perform the detailed computations (3.5.2), (3.5.5), (3.5.10), (3.5.11), (3.5.4) concerning their own products.

Let us examine the economic interpretation of the center's problem (3.5.14). Note that since $h^{\sigma-1}_{ij} = \Sigma_g\, d^{\sigma-1}_{gi} d^{\sigma-1}_{gj}$ and $b^{\sigma-1}_{j} = \Sigma_g\, d^{\sigma-1}_{gi} \hat{b}_g$, the center's problem can be restated as

$$\Sigma_g\, d^{\sigma-1}_{gi} (\Sigma_j\, d^{\sigma-1}_{gj} z_j - \hat{b}_g) = 0, \qquad i = 1, \ldots, J. \tag{3.5.15}$$

It can be seen that the ith equation of this problem has been obtained from the semiaggregate model (3.5.1) by adding together all equations in the semiaggregate model with the coefficients $d^{\sigma-1}_{gi}$, which indicate the importance given to the balance of product g in the input-output model of aggregate product i (if the coefficient $d^{\sigma-1}_{gi}$ is large in absolute value, it means that the importance of the balance of the product g is high). Accordingly, the solution of the aggregate model (3.5.15) will ensure a greater degree of balance (i.e., smaller imbalances) to those products g of the semiaggregate model (3.5.1) with larger values of $d_{gi}^{\sigma-1}$.

The most rational information flow scheme for the process (3.5.2), (3.5.5), (3.5.10)-(3.5.14), (3.5.4) is as follows. After performing the computations (3.5.2) each industry j passes to each industry i the numbers $x_{gj}^{\sigma-1}$, $g \in G_i$; that is, the industries pass to each other the (J - 1)G numbers $x_{gj}^{\sigma-1}$. For the computations (3.5.12)-(3.5.13) the jth industry passes to the center the J(J + 1)/2 numbers $h_{ij\tau}^{\sigma-1}$ as well as the J numbers $b_{i\tau}^{\sigma-1}$. Thus at this stage the industries pass to the center $J^2(J + 3)/2$ numbers. After the computation of (3.5.14) the center passes to the industries the vector z^σ; that is, it passes J^2 numbers in all. Thus in one iteration of the algorithm the industries pass (J - 1)G numbers among themselves and $J^2(J + 5)/2$ numbers are passed in total between the industries and the center and back again.

In comparison with the algorithm (3.2.9)-(3.2.11), the industries pass the same amount of information among themselves, while the amount being passed from the industries to the center increases by $[J^2(J + 3)/2] - J$. If J is considerably smaller than G (which is usually the case), this increase in the volume of information passed between the subsystems is insignificant in comparison to the total amount of information handled by the system.

It is often quite difficult to estimate the Euclidean norm of the matrix A in equation (3.1.1). It is much easier to work with the ℓ_1 norm of matrix A, which is equal to $\max_q \Sigma_g |\hat{a}_{gq}|$. (Note also that the productivity condition (3.1.3) happens to be very close to the condition that the ℓ_1 norm of A is less than 1 but not always equivalent.) This is why it would seem natural to modify the process (3.5.2)-(3.5.4) in order to calculate z_i minimizing the ℓ_1 norm of the imbalances vector $D^{\sigma-1}z^\sigma - b$. In other words, this process would differ from (3.5.2)-(3.5.4) in that we replace (3.5.3) by solving the problem

$$\min_z |\Delta| = \min_z \sum_i \sum_{g \in G_i} \left| \sum_j x_{gj}^{\sigma-1} z_j + \hat{b}_g - x_g^{\sigma-1} z_i \right|.$$

Of course, it is desirable to find z^σ that makes the ℓ_1 norm as small as possible, but unlike the case of the Euclidean norm, this is a problem of considerable difficulty. Therefore, it seems reasonable to approximate at each iteration with a certain degree of accuracy the vector z^σ that minimizes the sum of the absolute imbalances at each iteration. This degree of accuracy will be based on a compromise between our desire to find the minimizing z^σ and our desire to perform a relatively small number of calculations.

The function $F(z) = \Sigma_g |\Sigma_i d_{gi}^{\sigma-1} z_i - \hat{b}_g|$ is convex, and due to this we can apply the generalized gradient method to search for its minimum although its derivatives are not defined everywhere (see, e.g., Shor, 1968); note that in this case the generalized gradient (subdifferential) of F(z) is easily defined[1] (see Rockafellar, 1970). Note also that grad $F(z) = \Sigma_j$ grad $F_j(z)$, where $F_j(z) = \Sigma_{g \in G_j} |\Sigma_i d_{gi}^{\sigma-1} z_i - \hat{b}_g|$.

Hence each subsystem can compute its own share of the gradient and pass it to the center, which will add them up to obtain the full gradient. Then the center sends the full gradient to the subsystems so that they could calculate their own values of steps, from which the center chooses the smallest one. Using this step size, the center recalculates the approximation of z. In fulfilling such procedures the center calculates only aggregated indices and the industries calculate detailed indices. Thus when minimizing F(z) by means of the gradient method the center calculates the sequence $z^{0,\sigma}, z^{1,\sigma}, \ldots, z^{\gamma,\sigma}$ such that $F(z^{\gamma,\sigma}) < F(z^{\gamma-1,\sigma})$. At each iteration the vector $z^{0,\sigma}$ is defined as a vector (1,...,1).

The value of $z^{1,\sigma}$ is most easily computed when all $\Delta_g^{0,\sigma-1} = \Sigma_i d_{gi}^{\sigma-1} z_i^{0,\sigma} - \hat{b}_g$ are not zero for g = 1, ..., G [due to (3.2.11),

[1]The problem of minimization of F(z) can be also transformed to a linear programming problem.

this will almost always be true]. In this instance,[1]

$$\operatorname{grad} F_j(z^{0,\sigma}) = \left\{ \sum_{g \in G_j} d_{gi}^{\sigma-1} \operatorname{sign} \Delta_g^{0,\sigma-1},\ i = 1, \ldots, J \right\}$$

and therefore

$$\operatorname{grad} F(z^{0,\sigma}) = \{\Sigma_g\ d_{gj}^{\sigma-1} \operatorname{sign} \Delta_g^{0,\sigma-1},\ i = 1, \ldots, J\}.$$

Let $\Sigma_g\ (\operatorname{sign} \Delta_g^{0,\sigma-1}) d_{gi}^{\sigma-1}$ be denoted by $\Delta z_i^{0,\sigma-1}$. Then we find

$$\alpha^1 = \min_g \frac{\Delta_g^{0,\sigma-1}}{\Sigma_i\ d_{gi}^{\sigma-1} \Delta z_i^{0,\sigma}}$$

and compute $z^{1,\sigma}$ from the formulas

$$z_i^{1,\sigma} = 1 - \alpha^1 \Delta z_i^{0,\sigma}, \qquad i = 1, \ldots, J.$$

A similar procedure could be used to calculate $z^{\gamma,\sigma}$, $\gamma > 1$, but when we have zero values of $\Delta_g^{\gamma,\sigma-1}$, the procedure could be more complicated. Therefore, taking z_i^{σ} equal to $z_i^{1,\sigma}$ would be a reasonable approach.

When z_g^{σ} is calculated by performing one step of the gradient minimization algorithm followed by one iteration of successive approximations at the second stage, one iteration of the entire algorithm proceeds as follows:

1. Compute the semiaggregate input flows:

$$d_{gj} = \begin{cases} \Sigma_{q \in G_j}\ \hat{a}_{gq} x_q^{\sigma-1}, & g \notin G_j; \\ \Sigma_{q \in G_j}\ \hat{a}_{gq} x_q^{\sigma-1} - x_g^{\sigma-1}, & g \in G_j. \end{cases} \tag{3.5.16}$$

---

[1]Remember that

$$\operatorname{sign} a = \begin{cases} 1 & \text{if } a > 0, \\ 0 & \text{if } a = 0, \\ 1 & \text{if } a < 0. \end{cases}$$

2. Calculate the vector of imbalances:

$$\Delta_g^{\sigma-1} = \Sigma_i\ d_{gi}^{\sigma-1} - \hat{b}_g, \qquad g = 1, \ldots, G. \tag{3.5.17}$$

3. Calculate the direction of the step:

$$\Delta z_i^{\sigma} = \Sigma_g\ \mathrm{sign}(\Delta_g^{\sigma-1}) d_{gi}^{\sigma-1}. \tag{3.5.18}$$

4. Determine the step size in that direction:

$$\alpha^{\sigma} = \min \frac{\Delta_g^{\sigma-1}}{\Sigma_i\ d_{gi}^{\sigma-1} \Delta z_i^{\sigma}}. \tag{3.5.19}$$

5. Calculate the aggregate multipliers:

$$z_i^{\sigma} = 1 - \alpha^{\sigma} \Delta z_i^{\sigma}, \qquad i = 1, \ldots, J. \tag{3.5.20}$$

6. Calculate the next approximation:

$$x_g^{\sigma} = \Sigma_j\ \hat{a}_{gq} x_q^{\sigma-1} z_j^{\sigma} + \hat{b}_g, \qquad 1, \ldots, G. \tag{3.5.21}$$

3.6 GENERALIZATIONS OF THE BASIC ALGORITHMS; THEIR VARIATIONS AND MODIFICATIONS

One iteration of the basic iterative aggregation algorithm (3.2.9)-(3.2.11) can be seen to consist of the following steps. First, the output proportions from the previous iteration are adjusted based on the industry outputs obtained from the input-output model. Then we perform one step of successive approximation on these adjusted values.

The first step consists of three distinct substeps: the computation of the aggregate indices (3.2.9), solution of the aggregate input-output model (3.2.10), and proportionally adjusting the outputs of the specific products based on the volume changes at the industry level:

$$\tilde{x}_q^{\sigma} = z_j^{\sigma} x_q^{\sigma-1}, \qquad q \in G_j, \qquad j = 1, \ldots, J.$$

The second step entails calculating one iteration of successive approximations with values $\tilde{x}_q^\sigma$:

$$x_g^\sigma = \Sigma_q \hat{a}_{gq}\tilde{x}_q^\sigma + \hat{b}_g, \qquad g = 1, \ldots, G.$$

Using the notation above, let us repeat the first basic two-level iterative aggregation process:

1. Compute aggregated flows:

$$X_i^{\sigma-1} = \sum_{g\in G_i} x_g^{\sigma-1}; \qquad X_{ij}^{\sigma-1} = \sum_{g\in G_i}\sum_{q\in G_j} \hat{a}_{gq}x_q^{\sigma-1}. \tag{3.6.1}$$

2. Solve the interindustry input-output model:

$$z_i X_i^{\sigma-1} = \sum_j z_j X_{ij}^{\sigma-1} + b_i, \qquad i = 1, \ldots, J, \qquad b_i = \sum_{g\in G_i} \hat{b}_g. \tag{3.6.2}$$

and denote the solution by z_i^σ.

3. Compute the intermediate solution:

$$\tilde{x}_g^\sigma = z_i^\sigma x_g^{\sigma-1}, \qquad g \in G_i. \tag{3.6.3}$$

4. Compute the new approximation:

$$x_g^\sigma = \sum_j \sum_{q\in G_j} \hat{a}_{gq}\tilde{x}_q^\sigma + \hat{b}_g. \tag{3.6.4}$$

Dudkin and Kasparson (1979) point out that we can construct a more general iterative aggregation process, which we shall refer to as a generalized iterative aggregation process. Let $y = Dx + c$ be an arbitrary linear transformation satisfying the properties $D \geq 0$, $C > 0$, and $\|D\| < 1$, and we assume that the solution x^* of the original problem $x = Ax + b$ is a stationary point of the transformation, that is, $x^* = Dx^* + C$. Then proceed as in the basic process: Construct an aggregate system (3.6.2) and compute the multipliers z_i^σ. Then we calculate the vector $\tilde{x}^\sigma = \{\tilde{x}_g^\sigma\}$ by the formulas

$$\tilde{x}_g^\sigma = \{z_i^\sigma x_g^{\sigma-1}\}, \qquad g \in G_i.$$

Next, instead of calculating one iteration of successive approximations, we calculate the transformation $y = Dx + c$, where the matrix D is not necessarily equal to the matrix A. That is, the vector x^σ, which is the approximation determined at the end of iteration σ, is calculated as follows:

$$x^\sigma = D\tilde{x}^\sigma + C. \qquad (3.6.5)$$

The first basic iterative aggregation process is derived from the generalized process by setting $D = A$ and $C = b$. If we use a method of solving linear systems that differs from successive approximation step (3.6.4) of the algorithm, we can generate other variants of the two-level iterative aggregation algorithm. All the variants represent different particular cases of the generalized algorithm (3.6.1)-(3.6.3), (3.6.5) with assumed conditions for D and C.

If we use at each iteration the so-called simple successive approximation process, the vector $x^\sigma = \{x_g^\sigma,\ g = 1, \ldots, G\}$ is calculated by the formulas

$$x_g^\sigma = \frac{1}{1 - a_{gg}} \sum_{q \neq g} \hat{a}_{gq}\tilde{x}_q^\sigma + \hat{b}_g, \qquad g = 1, \ldots, G. \qquad (3.6.6)$$

If we use at each iteration the Gauss-Seidel process, the vector $x^\sigma = \{x_g^\sigma,\ g = 1, \ldots, G\}$ is determined by the formulas

$$\begin{aligned} x_1^\sigma &= \sum_{q=1}^{G} \hat{a}_{1q}\tilde{x}_q^\sigma + \hat{b}_1; \\ x_2^\sigma &= \hat{a}_{21}x_1^\sigma + \sum_{q=2}^{G} \hat{a}_{gq}\tilde{x}_q^\sigma + \hat{b}_2; \qquad (3.6.7) \\ x_3^\sigma &= \hat{a}_{31}x_1^\sigma + \hat{a}_{32}x_2^\sigma + \sum_{q=3}^{G} \hat{a}_{gq}\tilde{x}_q^\sigma + \hat{b}_3; \qquad \text{etc.} \end{aligned}$$

Using one step of Nekrasov's process yields

$$\begin{aligned} x_1^\sigma &= \frac{1}{1 - \hat{a}_{11}} \left(\sum_{q=2}^{G} \hat{a}_{1q}\tilde{x}_q^\sigma + \hat{b}_1 \right); \\ x_2^\sigma &= \frac{1}{1 - \hat{a}_{22}} \left(\hat{a}_{21}x_1^\sigma + \sum_{q=3}^{G} \hat{a}_{2q}\tilde{x}_q^\sigma + \hat{b}_2 \right); \end{aligned} \qquad (3.6.8)$$

$$x_3^\sigma = \frac{1}{1 - \hat{a}_{33}}\left(\hat{a}_{31}x_1^\sigma + \hat{a}_{32}x_2^\sigma + \sum_{q=4}^{G} a_{3q}\tilde{x}_q^\sigma + \hat{b}_3\right); \quad \text{etc.} \qquad (3.6.8)\ \text{cont.}$$

Other variants of the two-level iterative aggregation process can be constructed by using other methods for solving the linear system at the second substep (3.6.2) of the aggregation-disaggregation step (3.6.1)-(3.6.3) of the generalized process (3.6.1)-(3.6.3), (3.6.6). Three further variants can be found by using the generalized versions[1] of the three methods just described for solving linear systems. In these additional processes the matrix in the original system is partitioned into blocks that correspond to industries.

When we realize the second step (3.6.4) in the generalized form of the first method, we solve a linear system in the variables of that block with the variables of the other blocks fixed at the values that we have obtained from the third substep (3.6.3) of the given iteration. In other words,

$$x_g = \sum_{q \in G_i} \hat{a}_{gq}x_q + \sum_{q \notin G_i} \hat{a}_{gq}\tilde{x}_q^\sigma + \hat{b}_g, \qquad g \in G_i,\ i = 1, \ldots, J. \qquad (3.6.9)$$

We can call this method the block simple successive approximation method.

The generalized form of the second method (i.e., block Gauss-Seidel method) obtains the solution by determining the new values of the variables in each block successively. That is, we use the new fixed values of variables of blocks already calculated at the given iteration, and the values that we obtained from substep (3.6.3) of the given iteration for blocks yet to be calculated:

$$\begin{aligned} x_g^\sigma &= \Sigma_q\, \hat{a}_{gq}\tilde{x}_q^\sigma + b_g, \qquad g \in G_1 \\ x_g^\sigma &= \sum_{q \in G_1} \hat{a}_{gq}x_q^\sigma + \sum_{q \notin G_1} \hat{a}_{gq}\tilde{x}_q^\sigma + \hat{b}_g, \qquad g \in G_2; \end{aligned} \qquad (3.6.10)$$

---

[1]We must pay attention that these generalized versions are generalizations of the three processes just considered, but they are at the same time only particular cases of the generalized process (3.6.1)-(3.6.3), (3.6.5).

$$x_g^\sigma = \sum_{q \in G_1, G_2} \hat{a}_{gq} x_q^\sigma + \sum_{q \notin G_1, G_2} \hat{a}_{gq} \tilde{x}_q^\sigma + \hat{b}_g, \qquad g \in G_3; \qquad (3.6.10)$$
cont.

etc.

The second step of the third generalized form is a combination of (3.6.9), (3.6.10), that is, we solve successively the following systems:

$$x_g = \sum_{q \in G_1} \hat{a}_{gq} x_q + \sum_{q \notin G_1} \hat{a}_{gq} \tilde{x}_q^\sigma + \hat{b}_g, \qquad g \in G_1;$$

$$x_g = \sum_{q \in G_1} \hat{a}_{gq} x_q^\sigma + \sum_{q \in G_2} \hat{a}_{gq} x_q + \sum_{q \notin G_1, G_2} a_{gq} \tilde{x}_q^\sigma + \hat{b}_g, \qquad g \in G_2; \qquad (3.6.11)$$

etc.

We can also construct a generalized version of the classical (basic) one-level process (3.3.5), (3.3.6). Assume that the equation

$$x = Dx + C, \qquad D = \{d_{gq}\}, \qquad C = \{c_g\}, \qquad d_{gq} \geq 0, \qquad c_g > 0$$

has the same solution as the original problem. Then in the generalized one-level process formulas (3.3.6) are replaced by the formulas

$$x_g = \sum_{q \in G_j} d_{gq} x_q + \sum_{\tau \neq j} z_\tau^j \sum_{q \in G_\tau} d_{gq} x_q^{\sigma - 1} + c_g, \qquad g \in G_j,$$

$$z_i^j X_i^{\sigma-1} = \sum_{q \in G_j} \sum_{g \in G_i} \hat{a}_{gq} x_q + \sum_{\tau \neq j} z_\tau^j X_{i\tau}^{\sigma-1} + b_i, \qquad (3.6.12)$$

$$i = 1, \ldots, j-1, j+1, \ldots, J.$$

The process (3.3.5), (3.3.6) does not have any further variations since the variables x_g and z_i^j are computed by solving one problem (3.3.6).

However, we can construct a one-level process in which the two sets of variables are derived from different problems, for example, from two problems solved successively for each industry at each iteration, an interindustry input-output problem and then an intraindustry problem. In the generalized form of this process, the two problems solved by the industries have the following form:

$$z_i^{j,\sigma}X_i^{\sigma-1} = \sum_{g\in G_i} \sum_{q\in G_j} \hat{a}_{gq}x_q^{\sigma-1} + \sum_{\tau\neq j} z_\tau^{j,\sigma}X_{i\tau}^{\sigma-1} + b_i, \qquad (3.6.13)$$

$$i = 1, \ldots, j-1, j+1, \ldots, J;$$

$$x_g^\sigma = \sum_{q\in G_j} d_{gq}x_q^{\sigma-1} + \sum_{\tau\neq j} z_\tau^{j,\sigma} \sum_{q\in G_\tau} d_{gq}x_q^{\sigma-1} + c_g, \qquad (3.6.14)$$

$$g \in G_j.$$

The different variants for this one-level process with two industry problems are obtained by the appropriate choice of the matrix D and the vector C as in the description of the variants of the two-level processes.

Appendix 3.4 contains proofs of convergence for the generalized processes subject to certain constraints on matrices A and D. [A short version of these proofs appeared in the paper by Dudkin and Kasparson (1979).] These results state that convergence is guaranteed for all the variants of the generalized processes as long as the norm of the original matrix A is small enough. However, numerical experiments suggest that these algorithms converge under the general condition that $\|A\| < 1$.

The two-level block processes have a structure similar to the basic one-level process (3.3.5), (3.3.6). In the block processes, after we have solved the upper-level problem for the multipliers z_i^σ, each industry solves an independent problem in order to find the specific outputs x_g^σ, $g \in G_i$. Numerical experiments suggest that the block algorithms have a rate of convergence similar to that found for the basic one-level algorithm. Conversely, the variants of the one-level problem that have each industry solve a separate intra- and interindustry problem do not demonstrate any increase in the rate of convergence. As these methods require more computations per iteration, their study appears to be of only theoretical interest.

We can use the same ideas to generate variants of the algorithms described in Section 3.5. In Appendixes 3.2 and 3.3 we prove that if the norm of the matrix in the original system is small enough, all the variants will converge. Also, for the generalized two-level algorithm we can replace the linear transformation Dx + C by any other, not necessarily linear transformation $L(x) \geqq 0$ with the following properties: $L(x) \geqq 0$ for $x \geqq 0$, $L(x^*) = x^*$, and for $x \geqq 0$, $\|L(x) - x^*\| \leqq \beta\|x - x^*\|$, where $\beta < 1$.

There is one other class of iterative aggregation algorithms that can be constructed based on (3.6.1)-(3.6.4) and (3.3.5), (3.3.6). In this class of algorithms we calculate approximate rather than exact solutions to the aggregate systems. Then using the approximate solution we compute the new approximation by formulas (3.6.4) or (3.3.6). Algorithms of this type were examined in Vakhutinsky and Kasparson (1977), where they were called modified iterative aggregation processes. We describe two of these modified algorithms for the two-level iterative aggregation process.

Modification I. An approximate solution of the aggregate system (3.6.2) is found by performing τ steps of successive approximations with initial values $z_i^{0,\sigma} = 1$; that is, the values $z_i^{\tau,\sigma}$ are successively determined from the equations

$$X_i^{\sigma-1} z_i^{\tau,\sigma} = \Sigma_j\, z_j^{\tau-1,\sigma} X_{ij}^{\sigma-1} + \tilde{b}_i; \qquad z_j^{0,\sigma} = 1. \tag{3.6.15a}$$

We compute the new approximation by the formulas

$$x_g^{\sigma} = \sum_j z_j^{\tau,\sigma} \sum_{q\in G_j} \hat{a}_{gq} x_q^{\sigma-1} + \hat{b}_g. \tag{3.6.15b}$$

In Appendix 3.4 it is proven that if $\|A\| < 1/3$, this algorithm will converge to the solution of the original system. Numerical experiments showed that this process converges as long as $\|A\| < 1$ and converges at a rate that is no slower than that of the original process (3.6.1)-(3.6.4), even when $\tau = 1$; that is, only one step of successive approximations is calculated at each iteration. As one iteration of this modified process requires fewer computations than the original process, it is certainly of more than just theoretical interest.

Modification II. In this modification any method can be used to solve the original system approximately, but the degree of approximation needed will depend on the degree of approximation of the original system at iteration $(\sigma - 1)$. To do this we calculate the norm of the imbalances vector

$$\Delta^{\sigma} = \sum_g \left|\sum_q \hat{a}_{gq} x_q^{\sigma-1} - x_g^{\sigma-1} - b_g\right|.$$

Then by an arbitrary method we solve the aggregate system (3.6.2) to an accuracy of $k\Delta^{\sigma}$ (k being a given number). It means that we find $\tilde{z}_i^{\sigma}$ not subject to the system of equations (3.6.2), but to the inequality

$$\sum_i \left|\tilde{z}_i^{\sigma} X_i^{\sigma-1} - \sum_j \tilde{z}_j^{\sigma} X_{ij}^{\sigma-1} - \tilde{b}_i\right| \leq k\Delta^{\sigma}. \tag{3.6.16}$$

Using this solution, we compute the new approximation

$$x_g^{\sigma} = \sum_j \tilde{z}_j^{\sigma} \sum_{q \in G_j} \hat{a}_{gq} x_q^{\sigma-1} + \hat{b}_g. \tag{3.6.17}$$

The convergence of this process is also discussed in Appendix 3.4.

Modification II is not particularly useful for practical computations, but it can be used to form a theoretical basis for solving the aggregate system by approximate methods. Similar modifications can also be constructed for the one-level process (3.3.5), (3.3.6). One such modification was presented by Kasparson (1976a,b).

One other possible generalization of the basic two-level process is as follows. If we have the problem

$$x_g = \Sigma_q \hat{a}_{gq} x_q + \hat{b}_g, \qquad g = 1, \ldots, G,$$

and the matrix

$$C^{\sigma-1} = \{c_{iq}^{\sigma-1}\}, \qquad i = 1, \ldots, J, \qquad q = 1, \ldots, G,$$

of rank J, we can take as our master problem the generalized input-output model

$$\sum_j z_j \sum_{q \in G_j} c_{iq}^{\sigma-1} x_q^{\sigma-1} = \sum_j z_j \sum_{g \in G_i} \sum_{q \in G_j} c_{ig}^{\sigma-1} \hat{a}_{gq} x_q^{\sigma-1} + \sum_{g \in G_i} \hat{b}_g$$

or in matrix form,

$$C^{\sigma-1}X^{\sigma-1}z = C^{\sigma-1}AX^{\sigma-1}z + C^{\sigma-1}b. \qquad (3.6.18)$$

Here $X^{\sigma-1} = \{x_{gi}^{\sigma-1}\}$, $g = 1, \ldots, G$, $i = 1, \ldots, J$, $x_{gi}^{\sigma-1} = x_g^{\sigma-1}$ if $g \in G_i$ and $x_{gi}^{\sigma-1} = 0$ if $g \notin G_i$.

The system of equations (3.6.2) can be derived from (3.6.18) by setting $c_{ig}^{\sigma-1} = 1$ if $g \in G_i$ and $c_{ig}^{\sigma-1} = 0$ if $g \notin G_i$. We can also use other methods of choosing the matrix $C^{\sigma-1}$. For example, at iteration $\sigma - 1$ we can partition the constraints into J groups such that the sum of the imbalances in each group will be approximately equal. Let M_j index the constraints in the jth group and set $c_{ig}^{\sigma-1} = 1$ if $g \in M_i$, $c_{ig}^{\sigma-1} = 0$ if $g \notin M_i$. Another method of choosing $C^{\sigma-1}$ consists in introducing the quantities $\varepsilon_g^{\sigma-1} = |\Sigma_q \hat{a}_{gq} x_q^{\sigma-1} - \hat{b}_g|$ and setting the $c_{ig}^{\sigma-1}$ equal to $\varepsilon_g^{\sigma-1}/\Sigma_{g\in G_i} \varepsilon_g^{\sigma-1}$ if $g \in G_i$, or to 0 otherwise. This scheme generates the algorithm (3.5.2), (3.5.3), (3.5.4) if

$$c_{ig}^{\sigma-1} = \delta_{ij} x_g^{\sigma-1} - \sum_{q \in G_i} \hat{a}_{gq} x_q^{\sigma-1}, \qquad g \in G_j;$$

$$\delta_{ij} = 1 \text{ if } i = j; \qquad \delta_{ij} = 0 \text{ if } i \neq j.$$

Ven and Erlich (1970) demonstrate how to determine $C^{\sigma-1}$ such that the solution z^σ of (3.6.18) would be equal to $C^{\sigma-1}x^*$, where x^* is the solution of the original system. They have also shown how to determine x^* from such a z^σ. Thus they have developed a new finite method of solving (3.1.1). However, this method appears to be computationally very costly. It is quite natural, though, to use (3.6.18) to adjust various methods of solving (3.1.1).

3.7 SIMPLEST CONVERSION OF ITERATIVE AGGREGATION ALGORITHMS FOR INTERPRODUCT INPUT-OUTPUT MODELS INTO PROVABLY CONVERGENT ALGORITHMS

Dudkin and Kasparson (1983) have shown that the basic two-level iterative aggregation process as well as its variations and modifications described in Section 3.6 can be transformed in such a way that they can be proven to converge. Their idea is to modify step (3.6.2) in the process (3.6.1)-(3.6.4). On this step the values of z_i^σ are found, and we consider other algorithms for choosing z_i^σ. Note first that if we always take $z_i^\sigma = 1$, we always get a successive approximation method that converges. Moreover, we can generalize this statement in the following way.

THEOREM 3.7.1 Consider the process (3.6.3), (3.6.4) with any algorithm to choose z_i^σ. Then the following condition is sufficient for it to converge to the solution x* of the problem (3.1.1):

$$\lim_{\sigma\to\infty} z_i^\sigma = 1. \tag{3.7.1}$$

Proof: The process can be stated as

$$x^\sigma = A^\sigma x^{\sigma-1} + b, \tag{3.7.2}$$

where the matrix $A^\sigma = (\hat{a}_{gq}^\sigma)$ has the form $\hat{a}_{gq}^\sigma = \hat{a}_{gq} z_j^\sigma$, $g \in G_j$. Let $x^\sigma = x^* + \varepsilon^\sigma$, $A^\sigma = A + \Lambda^\sigma$. Then (3.7.2) can be written as

$$\varepsilon^{\sigma+1} = (A + \Lambda^{\sigma+1})\varepsilon^\sigma + \Lambda^{\sigma+1}x^*. \tag{3.7.3}$$

From the assumption that $z_j^\sigma \to 1$, it follows that $\Lambda^\sigma \to 0$. Then, starting from some σ we have

$$\|A + \Lambda^\sigma\| \le \beta < 1 \tag{3.7.4}$$

and

$$\|\Lambda^\sigma x^*\| = m_\sigma \to 0. \tag{3.7.5}$$

From (3.7.4) we have

$$\|(A + \Lambda^{\sigma+1})\varepsilon^\sigma\| \le \|A + \Lambda^{\sigma+1}\|\cdot\|\varepsilon^\sigma\| \le \beta\|\varepsilon^\sigma\|. \tag{3.7.6}$$

From (3.7.3), (3.7.5), (3.7.6) we obtain the estimate

$$\begin{aligned}\|\varepsilon\|^{\sigma+1} &= \|(A + \Lambda^{\sigma+1})\varepsilon^\sigma + \Lambda^{\sigma+1}x^*\| \\ &\le \|(A + \Lambda^{\sigma+1})\varepsilon^\sigma\| + \|\Lambda^{\sigma+1}x^*\| \le \beta\|\varepsilon^\sigma\| + m_\sigma .\end{aligned}$$

Next we find

$$\|\varepsilon^{\sigma+1}\| \leq \beta^{\sigma+1}\|\varepsilon^0\| + \sum_{k=0}^{\sigma} \beta^k m_{\sigma-k}.$$

Thus, to prove convergence, it is sufficient to prove that

$$\lim_{\sigma\to\infty} \sum_{k=0}^{\sigma} \beta^k m_{\sigma-k} = 0.$$

For an arbitrary $\delta > 0$, choose σ' such that for any $\sigma \geqq \sigma'$ the inequalities $m_\sigma < \delta$ and $\beta^\sigma < \delta$ are satisfied. Then for any σ such that $\sigma \geqq 2\sigma'$ we have

$$\sum_{k=0}^{\sigma} \beta^k m_{\sigma-k} = \sum_{k=0}^{\sigma'} \beta^k m_{\sigma-k} + \sum_{k=\sigma'+1}^{\sigma} \beta^k m_{\sigma-k} < \frac{\delta}{1-\beta} + \frac{\delta M}{1-\beta},$$

where $M = \max m_\sigma < \infty$. As δ was chosen arbitrarily, the convergence of $\Sigma_{k=0}^{\sigma} \beta^k m_{\sigma-k}\sigma$ has been proven, and hence the convergence of the algorithm.

A similar proposition holds for the system $x = Bx + c$, where B is an arbitrary matrix whose eigenvalues are less than 1 in absolute value and where c is an arbitrary vector. The proof in this case is essentially the same as the proof of the previous proposition, differing only in nonessential technical details.

This approach enables us to convert the two-level iterative aggregation algorithm into a convergent process. To do this it is sufficient in the disaggregation step to replace the multipliers z_j^σ obtained from the solution of the aggregate problem with the values

$$\tilde{z}_j^\sigma = 1 + (z_j^\sigma - 1)q_\sigma,$$

where q_σ is an arbitrary sequence that approaches 0 in the limit. It is easy to show that for all such processes z_j^σ approaches 1. Thus the adjusted process converges if the classical iterative method of successive approximations when used to solve (3.1.1)-(3.1.3) converges. Moreover, the adjusted basic two-level process and its modifications converge, as do the variants of this process if the corresponding classical methods when used to solve (3.1.1)-(3.1.3) converge.

When performing numerical experiments, we used the sequence $q_\sigma = \lambda^\sigma$, where $0 \leqq \lambda < 1$, and the superscript on λ denotes a power of σ. When the value of λ was close to 1 during the initial iterations, the adjusted process behaved similarly to the corresponding unadjusted iterative aggregation process. However, the adjusted process would be a priori convergent. If λ were chosen to be close to 0, the adjusted processes behaved like the classical iterative methods. The fastest rates of convergence were found for λ with values in the range 0.85 to 0.90.

The imbalances-minimizing processes for semiaggregate problems can also be adjusted in a similar fashion. However, as these processes have already been shown to converge, the need for such adjusted processes is not clear.

APPENDIX 3.1 MATHEMATICAL ANALYSIS OF BASIC ITERATIVE AGGREGATION ALGORITHMS FOR INTERPRODUCT INPUT-OUTPUT MODELS

First we present the most significant of the current results concerning convergence of the basic two-level iterative aggregation algorithm for the model (3.1.1)-(3.1.3):

$$x_g = \Sigma_q \hat{a}_{gq} x_q + \hat{b}_g, \qquad g = 1, \ldots, G; \tag{3.1.1}$$

$$\hat{a}_{gq} \geq 0, \quad g, q = 1, \ldots, G; \qquad \hat{b}_g \geq 0, \quad g = 1, \ldots, G; \tag{3.1.2}$$

$$\max_q \Sigma_g \hat{a}_{gq} = \alpha < 1. \tag{3.1.3}$$

For J = 1 Shchennikov (1965) proved that the algorithm converges if the matrix $A = (\hat{a}_{gq})$ is nondecomposable. We prove the convergence of the algorithm for J = 1 if A is an arbitrary, not necessarily nondecomposable matrix.

THEOREM A3.1.1 If J = 1, the algorithm (3.2.5), (3.2.6), (3.2.3), (3.2.7) converges for any starting point x^0.

Proof: Rewrite the algorithm (3.2.5), (3.2.6), (3.2.3), (3.2.7) for J = 1.

1. Compute the aggregation weights:

$$d_g^\sigma = \frac{x_g^{\sigma-1}}{\Sigma_q x_q^{\sigma-1}}, \qquad g = 1, \ldots, G. \tag{A3.1.1}$$

2. Solve the aggregate input-input model:

$$X = X(\Sigma_g \Sigma_q \hat{a}_{gq} d_q^\sigma) + \Sigma_g \hat{b}_g \tag{A3.1.2}$$

   and denote the solution by X^σ.

3. Compute the next approximation:

$$x_g^\sigma = (\Sigma_q \hat{a}_{gq} d_q^\sigma) X^\sigma + \hat{b}_g. \tag{A3.1.3}$$

   Since

$$\Sigma_g d_g^\sigma = \frac{\Sigma_g x_g^{\sigma-1}}{\Sigma_g x_g^{\sigma-1}} = 1$$

we can write the solution of (A3.1.2) as

$$X^\sigma = \frac{\Sigma_g \hat{b}_g}{1 - \Sigma_g \Sigma_q \hat{a}_{gq} d_q^\sigma} = \frac{\Sigma_g \hat{b}_g}{\Sigma_q d_q^\sigma - \Sigma_q (\Sigma_g \hat{a}_{gq}) d_q^\sigma}$$

$$= \frac{\Sigma_g \hat{b}_g}{\Sigma_q (1 - \Sigma_g \hat{a}_{gq}) d_q^\sigma}. \tag{A3.1.4}$$

Adding G equations in (A3.1.3) yields

$$\Sigma_g x_g^\sigma = X^\sigma (\Sigma_g \Sigma_q \hat{a}_{gq} d_q^\sigma) + \Sigma_g \hat{b}_g = X^\sigma.$$

This identity, together with (A3.1.1), (A3.1.3), and (A3.1.4) implies that

$$d_g^{\sigma+1} = \frac{x_g^\sigma}{X^\sigma} = \frac{X^\sigma(\Sigma_q \hat{a}_{gq} d_q^\sigma) + \hat{b}_g}{X^\sigma}$$

$$= \Sigma_q \hat{a}_{gq} d_q^\sigma + \frac{\hat{b}_g}{\Sigma_h \hat{b}_h} \Sigma_q (1 - \Sigma_h \hat{a}_{hq}) d_q^\sigma$$

$$= \Sigma_q d_q^\sigma \left[\hat{a}_{gq} + \frac{\hat{b}_g}{\Sigma_h \hat{b}_h} (1 - \Sigma_h \hat{a}_{hq})\right]. \tag{A3.1.5}$$

Thus $d_g^{\sigma+1}$ is obtained by the transformation

$$d_g^{\sigma+1} = \Sigma_q \, s_{gq} d_q^{\sigma} \qquad \text{or} \qquad d^{\sigma+1} = Sd^{\sigma}.$$

Here S is a G × G matrix defined by

$$s_{gq} = \hat{a}_{gq} + \frac{\hat{b}_g}{\Sigma_h \hat{b}_h} (1 - \Sigma_h \hat{a}_{hq}). \tag{A3.1.6}$$

It is easy to see that $s_{gq} \geq 0$ and $\Sigma_g s_{gq} = 1$ for all q. Hence S is a stochastic matrix. We can see that if $\hat{b}_g \neq 0$, then also $s_{gq} \neq 0$. Without loss of generality we assume that $\hat{b}_g \neq 0$, $g = 1, \ldots, G_1$, $b_g = 0$, $g = G_1 + 1, \ldots, G$.

Consider the subscripts g contained in $G_1 + 1, \ldots, G$ for which there exist nonzero values of $\hat{a}_{gq}$, $q \leq G_1$. Without loss of generality we can assume that g is contained in $G_1 + 1, \ldots, G_2$. Next we can take those g in $G_2 + 1, \ldots, G$ for which $\hat{a}_{gq} > 0$, for some $q \leq G_2$, and so on. It is clear that for some $n \leq G$ the equality $G_n = G_{n+1}$ holds. Thus we can write the matrix S in the form

$$\begin{pmatrix} S_{11} & S_{12} \\ 0 & S_{22} \end{pmatrix},$$

where S_{11} is a stochastic nondecomposable matrix and S_{22} is a nonnegative matrix with norm less than 1. It follows from the Frobenius-Perron theorem that a nondecomposable stochastic N × N matrix has exactly one positive eigenvector with eigenvalue 1, and the remaining N - 1 eigenvalues are less than 1 in absolute value. Hence since

$$d_n^{\sigma+1} = S_{22} d_n^{\sigma}, \qquad d_n^{\sigma} = (d_g^{\sigma}, \ g > G_n),$$

this implies that

$$\lim_{\sigma\to\infty} d_g^{\sigma} = 0, \qquad g > G_n.$$

From this we conclude that

$$\lim d_g^\sigma, \qquad g \leq G_n,$$

exists and is greater than 0. This completes the proof of the theorem.

Now that we have shown that the algorithm converges for the case J = 1, let us turn to the rate of convergence of the algorithm.

LEMMA A3.1.1 If J = 1 and

$$\max_{q,h} \sum_g |\hat{a}_{gq} - \hat{a}_{gh}| \leq 3 \min_q \sum_g \hat{a}_{gq} - \max_q \sum_g \hat{a}_{gq},$$

the algorithm (3.2.5), (3.2.6), (3.2.3), (3.2.7) converges at a rate not slower than that of successive approximations.

Proof: Recall the following result by Lubich (1974). Let $S = (s_{gq})$ be a real matrix with column sums equal to 1. It is easy to see that the left multiplication by S maps the hyperplane $\{\Sigma_g x_g = 0\}$ into itself. Let δ_2 be the ℓ_1 norm of this map. Then

$$\delta_2 = \frac{1}{2} \max_{q,h} \sum_g |s_{gq} - s_{gh}|.$$

Applying this result to the stochastic matrix S of Theorem A3.1.1 yields

$$\begin{aligned}\delta_2 &= \frac{1}{2} \max_{q,h} \sum_g |s_{gq} - s_{gh}| = \frac{1}{2} \max_{q,h} \sum_g \Bigg|\hat{a}_{gq} - \hat{a}_{gh} \\ &\quad + \frac{\hat{b}_g}{\Sigma_\tau \hat{b}_\tau}\left(\sum_\tau \hat{a}_{\tau h} - \sum_\tau \hat{a}_{\tau q}\right)\Bigg| \leq \frac{1}{2} \max_{q,h} \sum_g |\hat{a}_{gq} - \hat{a}_{gh}| \\ &\quad + \frac{1}{2} \max_{q,h} \left|\sum_\tau \hat{a}_{\tau h} - \sum_\tau \hat{a}_{\tau q}\right| \frac{\Sigma_g \hat{b}_g}{\Sigma_\tau \hat{b}_\tau} \\ &\leq \frac{1}{2} \max_{q,h} \sum_g |\hat{a}_{gq} - \hat{a}_{gh}| + \frac{1}{2}\left(\max_q \sum_\tau \hat{a}_{\tau q} - \min_q \sum_\tau \hat{a}_{\tau q}\right).\end{aligned}$$

Note that δ_2 is greater than or equal to the second largest (in absolute value) eigenvalue of the matrix S, and the rate of convergence of the algorithm is determined by this eigenvalue. Thus for the rate of convergence to be not slower than that of successive approximations, it suffices to have $\delta_2 \leq \lambda_1$, where λ_1 is the largest (in absolute value) eigenvalue of the matrix A. Since for any nonnegative matrix A,

$$\lambda_1 \geq \min_q \sum_g \hat{a}_{gq},$$

it follows that $\delta_2 \leq \lambda_1$ whenever

$$\min_q \sum_g \hat{a}_{gq} \geq \frac{1}{2} \max_{q,h} \sum_g |\hat{a}_{gq} - \hat{a}_{gh}| + \frac{1}{2}\left(\max_q \sum_g \hat{a}_{gq} - \min_q \sum_g \hat{a}_{gq}\right).$$

The last inequality completes the proof of the lemma.

Note also that

$$\max_{q,h} \sum_g |\hat{a}_{gq} - \hat{a}_{gh}| \leq 2 \max_q \sum_g \hat{a}_{gq}.$$

Hence it follows from Lemma A3.3.1 that if the column sums of the matrix A are equal, the algorithm (3.2.5), (3.2.6), (3.2.3), (3.2.7) with J = 1 converges at a rate now slower than that of successive approximations. Also, if we are solving the problem (3.1.1)-(3.1.3), the quantity $1 - \Sigma_g \hat{a}_{gq}$ can be interpreted as the profit produced by 1 unit of product q. Therefore, if the profits produced by different products are approximately equal, the conditions of Lemma A3.1.1 should be satisfied in most cases.

It is easy to show that if J > 1 but G_j contains exactly one element for each j > 2, then

$$d^\sigma = \{d_g^\sigma\} = \left\{\frac{x_g^\sigma}{\Sigma_{q \in G_j} x_q^\sigma},\ g \in G_j\right\}$$

is a linear transformation of $d^{\sigma-1}$. Under these conditions and the additional assumption that the matrix A is nondecomposable, Shchennikov (1965) proved the convergence of the algorithm. However, as in Theorem A3.1.1, the nondecomposability assumption can be omitted.

The most general convergence result for the algorithm (3.2.5), (3.2.6), (3.2.3), (3.2.7) is given by the following theorem.

THEOREM A3.1.2 Let

$$\max_i \max_{q,h \in G_i} \sum_g |\hat{a}_{gq} - \hat{a}_{gh}| < 1 - \max_q \sum_g \hat{a}_{gq}. \tag{A3.1.7}$$

Then the algorithm (3.2.5), (3.2.6), (3.2.3), (3.2.7) monotonically converges in the ℓ_1 norm.

Proof: Without loss of generality we assume that

$$G_i = \{g | s_{i-1} \le g < s_i,\ 1 = s_0 < \cdots < s_J = G + 1\}.$$

Define the auxiliary matrices $T = (t_{jg})$ and $V^\sigma = (v^\sigma_{gj})$, $j = 1, \ldots, J$, $g = 1, \ldots, G$ by $t_{jq} = 1$ if $q \in G_j$, $t_{jq} = 0$ if $q \notin G_j$, and $v^\sigma_{gj} = d^\sigma_g$ if $g \in G_j$, $v^\sigma_{gj} = 0$ if $g \notin G_j$. Here

$$d^\sigma_g = \frac{x^{\sigma-1}_g}{\sum_{q \in G_j} x^{\sigma-1}_q}, \qquad g \in G_j.$$

As in the case of formulas (A3.1.4), it can be shown for any σ that

$$X^\sigma_j = \sum_{g \in G_j} x^\sigma_g \tag{A3.1.8}$$

or in matrix form that $X^\sigma = Tx^\sigma$. Using this result and the matrix V^σ defined above, we can write (3.2.7) as

$$x^\sigma = AV^\sigma X^\sigma + b = AV^\sigma(Tx^\sigma) + b = (AV^\sigma T)x^\sigma + b. \tag{A3.1.9}$$

Since

$$\|AV^\sigma T\| \le \|A\| \cdot \|V^\sigma T\|$$

and

$$\|V^\sigma T\| = 1,$$

the system (A3.1.9) is nondegenerate. Therefore,

$$x^\sigma = (I - AV^\sigma T)^{-1} b = (I - AV^\sigma T)^{-1}(I - A)x,$$

where x is the solution of the original system. Hence

$$\begin{aligned} x - x^\sigma &= [I - (I - AV^\sigma T)^{-1}(I - A)]x \\ &= [I - I - AV^\sigma T - (AV^\sigma T)^2 - \cdots + (I - AV^\sigma T)^{-1}A]x \\ &= [(I + AV^\sigma T + (AV^\sigma T)^2 + \cdots)(-AV^\sigma T) \\ &\quad + (I - AV^\sigma T)^{-1}A]x \\ &= [(I - AV^\sigma T)^{-1}(-AV^\sigma T) + (I - AV^\sigma T)^{-1}A]x \\ &= (I - AV^\sigma T)^{-1}(A - AV^\sigma T)x. \end{aligned} \tag{A3.1.10}$$

Since $V^\sigma Tx^{\sigma-1} = x^{\sigma-1}$, therefore

$$(A - AV^\sigma T)x^{\sigma-1} = A(x^{\sigma-1} - V^\sigma Tx^{\sigma-1}) = 0.$$

This implies that

$$(x - x^\sigma) = (I - AV^\sigma T)^{-1}(A - AV^\sigma T)(x - x^{\sigma-1}). \tag{A3.1.11}$$

Hence

$$\|x - x^\sigma\| \le \|(I - AV^\sigma T)^{-1}\|\|A - AV^\sigma T\|\|x - x^{\sigma-1}\|. \tag{A3.1.12}$$

The norm of the matrix $(I - AV^\sigma T)^{-1}$ can be approximated as

$$\begin{aligned} \|(I - AV^\sigma T)^{-1}\| &= \|(I + AV^\sigma T) + (AV^\sigma T)^2 + \cdots\| \\ &\leqq \|I\| + \|AV^\sigma T\| + \|AV^\sigma T\|^2 + \cdots \\ &\leqq 1 + \|A\|\cdot\|V^\sigma T\| + \|A\|^2\cdot\|V^\sigma T\|^2 + \cdots \\ &= 1 + \|A\| + \|A\|^2 + \cdots = \frac{1}{1 - \|A\|}. \end{aligned} \tag{A3.1.13}$$

Here we have used the fact that the norm of the matrix $V^\sigma T$ equals 1 for all σ.

Next we estimate the norm of the matrix $A - AV^\sigma T$ by

$$\begin{aligned} \|A - AV^\sigma T\| &= \max_j \max_{q\in G_j} \sum_g \left|\hat{a}_{gq} - \sum_{h\in G_j} \hat{a}_{gh} d_h^{\sigma-1}\right| \\ &= \max_j \max_{q\in G_j} \sum_g \left|\sum_{h\in G_j} d_h^{\sigma-1}\hat{a}_{gq} - \sum_{h\in G_j} \hat{a}_{gh} d_h^{\sigma-1}\right. \end{aligned}$$

$$\leq \max_j \max_{q \in G_j} \sum_{h \in G_j} d_h^\sigma \sum_g |\hat{a}_{gq} - \hat{a}_{gh}|$$

$$\leq \max_j \max_{q,h \in G_j} \sum_g |\hat{a}_{gq} - \hat{a}_{gh}|. \tag{A3.1.14}$$

We have used the fact that for all j and σ it is true that $\Sigma_{g \in G_j} d_g^\sigma = 1$. The proof of the theorem follows from (A3.1.12)-(A3.1.14).

Khizder (1971) proved convergence assuming that $\|A\| < 1/3$. Since

$$\max_j \max_{q,h \in G_j} \sum_g |\hat{a}_{gq} - \hat{a}_{gh}| \leq 2 \max_q \sum_g \hat{a}_{gq},$$

it is clear that this is a limiting case of condition (A3.1.7). Kalinina (1973) and Khomyakov (1973) have proven both local and global convergence for the algorithm assuming the condition

$$\max_i \sum_g (\max_{q \in G_i} \hat{a}_{gq} - \min_{q \in G_i} \hat{a}_{gq}) < 1 - \|A\|. \tag{A3.1.15}$$

It is clear that this condition is also a special case of (A3.1.7). Using (A3.1.11), it is also easy to prove convergence if for all i, j = 1, ..., J and for any q, h $\in G_j$ that

$$\sum_{g \in G_i} \hat{a}_{gq} = \sum_{g \in G_i} \hat{a}_{gh}. \tag{A3.1.16}$$

Moreover, if the matrix A is also nondecomposable, the rate of convergence will be faster than that for simple iterations.

We now analyze the algorithm (3.3.2)-(3.3.4). Rabinovich (1976) proved that this algorithm converges if (A3.1.1) holds and also if J = 2. Also, the algorithm converges if either (A3.1.16) is valid or if

$$\sum_j \max_q \sum_{g \in G_j} \hat{a}_{gq} < \frac{1}{3}. \tag{A3.1.17}$$

Finally, the following theorem has been proven for the algorithm (3.3.2)-(3.3.4).

THEOREM A3.1.3 Let

$$\max_j \max_{q,h\in G_j} \sum_g |\hat{a}_{gq} - \hat{a}_{gh}| < 1 - 2\|A\|. \tag{A3.1.18}$$

Then (3.3.2)-(3.3.4) converges monotonically in the ℓ_1 norm.

Proof: Define the matrix $W_i = (w^i_{gq})$, g, q = 1, ..., G, by

$$w^i_{gq} = \begin{cases} 1 & \text{if } g = q \in G_i, \\ 0 & \text{otherwise.} \end{cases}$$

The matrix T_i is derived from the matrix T of Theorem A3.1.2 by replacing the ith row of T by the matrix

$$C_i = \begin{pmatrix} \underbrace{0}_{\sum_{j<i}|G_j|} & \underbrace{\begin{matrix} 1 & & & 0 \\ & \cdot & & \\ & & \cdot & \\ 0 & & & 1 \end{matrix}}_{|G_i|} & \underbrace{0}_{\sum_{j>i}|G_j|} \end{pmatrix}.$$

Here $|G_j|$ denotes the cardinality of G_j and the matrix V^σ_i is obtained by replacing the ith column of V^σ in Theorem A3.1.2 with the matrix C'_i. Let $\hat{x}^\sigma_i = \{\hat{x}^\sigma_{gi}\}$ denote the vector generated by one iteration of (3.2.9)-(3.2.11) with initial value $x^{\sigma-1}$ and the partition $\{1,\ldots,G\} = \cup_{j\neq i} G_j \cup_{g\in G_i} G_g$ (G_g consists of one element g). It is clear that $x^\sigma_g = \hat{x}^\sigma_{gi}$, $g \in G_i$. From this we have that $\hat{x}^\sigma_i$ satisfies the system

$$\hat{x}^\sigma_i = AV^\sigma_i T_i \hat{x}^\sigma_i + b.$$

Combining this with (A3.1.10), we have

$$x - \hat{x}^\sigma_i = (I - AV^\sigma_i T_i)^{-1}(A - AV^\sigma_i T_i)x.$$

From the definition of $\hat{x}^\sigma_i$ it follows that the components of the vector $x - \hat{x}^\sigma_i$ whose indexes are contained in G_i are equal to $x_g - x^\sigma_g$ while the other components have a value of zero. Hence

$$x - x^\sigma = \sum_i W_i(x - \hat{x}^\sigma_i) = \sum_i W_i(I - AV^\sigma_i T_i)^{-1}(A - AV^\sigma_i T_i)(x - x^{\sigma-1}).$$

Therefore,

$$\|x - x^\sigma\| \le \|\Sigma_i W_i (I - AV_i^\sigma T_i)^{-1}(A - AV_i^\sigma T_i)\| \|x - x^{\sigma-1}\|.$$

For any matrix $B = (b_{gq})$, let $|B|$ denote the matrix that has $|b_{gq}|$ as its (g,q)th element. Combining this with the definition of the ℓ_1 norm of a matrix, we have

$$\begin{aligned} \|x - x^\sigma\| &\leqq \|\Sigma_i W_i |(I - AV_i^\sigma T_i)^{-1}| \cdot |A - AV_i^\sigma T_i|\| \cdot \|x - x^{\sigma-1}\| \\ &\leqq \|\Sigma_i |W_i (I - AV_i^\sigma T_i)^{-1}|\| \cdot \| |A - AV^\sigma T|\| \cdot \|x - x^{\sigma-1}\| \\ &= \|\Sigma_i W_i (I - AV_i^\sigma T_i)^{-1}\| \cdot \|A - AV^\sigma T\| \cdot \|x - x^{\sigma-1}\|. \end{aligned} \tag{A3.1.19}$$

Inequality (A3.1.19) follows from the fact that for all q not in G_i, the (g,q)th elements of the matrices $A - AV^\sigma T_i$ and $A - AV^\sigma T$ are identical. If q is contained in G_i, the qth column of the matrix $A - AV_i^\sigma T_i$ is identically zero. Hence each element of the matrix $A - AV^\sigma T$ has an absolute value no less than that of the corresponding element in the matrix $A - AV_i^\sigma T_i$.

Since the series $I + AV_i^\sigma T_i + \cdots$ converges for any i and its elements are nonnegative, these elements can be permuted. Hence

$$\begin{aligned} \|\Sigma_i W_i (I - AV_i^\sigma T_i)^{-1}\| &= \|\Sigma_i W_i (I + AV_i^\sigma T_i + \cdots)\| \\ &\leqq \|\Sigma_i W_i\| + \|\Sigma_i W_i AV_i^\sigma T_i\| \\ &\quad + \|\Sigma_i W_i (AV_i^\sigma T_i)^2\| + \cdots. \end{aligned}$$

Note that $\Sigma_i W_i = I$. Estimate $\|\Sigma_i W_i (AV_i^\sigma T_i)^k\|$.

Direct computation yields that every element of the matrix $AV_i^\sigma T_i$ is either equal to $\Sigma_{q \in G} \hat{a}_{gq} d_q^\sigma$ if q is in G_i or if not, then to $\hat{a}_{gq}$. Thus the matrix $A + AV^\sigma T$ is a positive matrix whose elements are no less than the corresponding elements of the matrix $AV_i^\sigma T_i$ for all i. Combining this fact with the fact that all the matrices in question are nonnegative, we have

$$\|\Sigma_i W_i (AV_i^\sigma T_i)^m\| \leq \|\Sigma_i W_i (A + AV^\sigma T)^m\|$$
$$= \|(A + AV^\sigma T)^m\| \leq (2\|A\|)^m,$$

and therefore

$$\|\Sigma_i W_i (I - AV_i^\sigma T_i)^{-1}\| \leq 1 + \|A\| + (2\|A\|)^2 + \cdots = \frac{1}{1 - 2\|A\|}.$$

Here we have used the fact that (A3.1.18) implies that $2\|A\| < 1$. Thus we have shown that

$$\|x - x^\sigma\| \leq \frac{\|A - AV^\sigma T\|}{1 - 2\|A\|} \|x - x^{\sigma-1}\|. \qquad \text{(A3.1.20)}$$

Theorem A3.1.3 then follows from the inequality (A3.1.14). It follows from this theorem that the algorithm (3.3.2)-(3.3.4) converges for $\|A\| < 1/4$.

APPENDIX 3.2 INVESTIGATION OF CONVERGENCE OF ITERATIVE AGGREGATION ALGORITHMS WITH MANOVE-WEITZMAN APPROXIMATIONS OF VON NEUMANN PRICES

In this appendix we discuss the convergence of the algorithms (3.4.5)-(3.4.9) and (3.4.15)-(3.4.18) when $J = 1$. We also investigate how the rate of convergence is related to the vector of prices.

LEMMA A3.2.1 If $J = 1$, the algorithm (3.4.5)-(3.4.9) converges to the solution of the system of equations $x = Ax + b$ for an arbitrary price vector p and an arbitrary productive matrix A.

Proof: Replace the matrix $A = (\hat{a}_{gq})$ by the matrix $p_g \hat{a}_{gq} p_q^{-1}$ and the vectors $x^\sigma = (x_g^\sigma)$ by $(p_g x_g^\sigma)$ and apply Lemma A3.1.1.

LEMMA A3.2.2 Let A be an arbitrary productive matrix, $\lambda_1, \lambda_2, \ldots, \lambda_G$ be its eigenvalues (which are the same as for A') so that $\lambda_1 > |\lambda_2| \geq |\lambda_3| \geq \cdots \geq |\lambda_G|$, and p be the eigenvector of A' associated with the eigenvalue λ_1 (i.e., $A'p = \lambda_1 p$). Then the rate of convergence of the algorithm (3.4.5)-(3.4.9) with the price vector p and $J = 1$ is determined by the second largest eigenvalue λ_2. That is, for every starting vector x^0 there exists $c > 0$ such that

$$\|x - x^*\| \leq c|\lambda_2|^\sigma.$$

Proof: At each aggregation step we find z^σ such that $(p, z^\sigma x^\sigma) = (p, Az^\sigma x^\sigma) + (p, b)$. Since $x^* = Ax^* + b$ we have $px^* = pAx^* + pb$. Therefore, $z^\sigma x^\sigma = x^* + y$, where y satisfies the equation $(p, y) = (p, Ay)$.

Since $A'p = \lambda_1 p$, $(p, Ay) = (A'p, y)$, and $\lambda_1 < 1$, we rewrite this equation as $(p, y) = 0$. Then $x^{\sigma-1} = Az^\sigma x^\sigma + b = Ax^* + b + Ay = x^* + Ay$. Ay also satisfies the equation $(p, Ay) = 0$.

Let $x^\sigma = x^* + y^\sigma$. Then for $\sigma \geq 1$ we have $(p, y^\sigma) = 0$ and $z^\sigma = 1$. Therefore, y^σ belongs to the hyperplane $H = \{y \mid (p, y) = 0\}$ and $y^{\sigma+1} = Ay^\sigma$.

The eigenvalues of the restriction of A onto H are $\lambda_2, \lambda_3, \ldots, \lambda_G$ and the rate of convergence of y^σ to 0 is determined by the largest eigenvalue λ_2. That is, for some constant c we have $\|y^\sigma\| \leq c|\lambda_2|^\sigma$. Since $y^\sigma = x^\sigma - x^*$, the proof of the lemma is completed.

In general, the rate of convergence of the algorithm (3.4.5)-(3.4.9) depends on the choice of the price vector p. It is not true that the fastest rate of convergence is always attained for the vector of von Neumann prices.

Consider, for example, the following two-dimensional problem:

$$x = \begin{pmatrix} \frac{1}{2} & \frac{1}{4} \\ \frac{1}{4} & \frac{1}{2} \end{pmatrix} x + \begin{pmatrix} 2 \\ 1 \end{pmatrix}.$$

Here

$$A = \begin{pmatrix} \frac{2}{4} & \frac{1}{4} \\ \frac{1}{4} & \frac{2}{4} \end{pmatrix}, \qquad b = \begin{pmatrix} 2 \\ 1 \end{pmatrix}.$$

The eigenvalues of A are 3/4 and 1/4. The vector of von Neumann prices is (1.1). The rate of convergence of the algorithm (3.4.5)-(3.4.9) with an arbitrary price vector (p_1, p_2) is determined by the number

$$M = \left| \frac{1}{2} - \frac{1}{4}\left(\frac{p_1 + 2p_2}{2p_1 + p_2}\right) \right|.$$

For the vector of von Neumann prices this number equals 1/4, which agrees with Lemma A3.2.2. However, the vector of von Neumann prices does not ensure the fastest rate of convergence. For example, if $(p_1, p_2) = (2,3)$, then $M = 3/14 < 1/4$. Nevertheless, Lemma A3.2.2 gives some explicit estimate for the rate of convergence that in general is better than the rate of convergence of simple iterations.

Let us now investigate the sensitivity of the algorithm (3.4.5)-(3.4.9) for $J = 1$ to small perturbations of the price vector p. In other words, suppose that instead of a constant price vector p, we have a sequence of price vectors p^σ, chosen by some other procedure and tending to a limit:

$$\lim_{\sigma\to\infty} p^\sigma = p.$$

Suppose that x^σ is the σth approximation to the solution x* of the equation $x = Ax + b$. To obtain the next approximation $x^{\sigma-1}$, we should perform the following operations:

1. Find z^σ such that

$$z^\sigma(p^\sigma, x^\sigma) = z^\sigma(p^\sigma, Ax^\sigma) + (p^\sigma, b). \tag{A3.2.1}$$

2. Find the next approximation $x^{\sigma-1}$ by the formula:

$$x^{\sigma+1} = z^\sigma Ax^\sigma + b. \tag{A3.2.2}$$

If we replace x^σ by its multiple mx^σ, then z will be replaced by $(1/m)z$ and $x^{\sigma+1}$ will remain the same. Therefore, if we replace $x^{\sigma+1}$ by its multiple $d^{\sigma+1}$ such that $(p^\sigma, x^\sigma) = (p^\sigma, d^{\sigma+1})$, it will not affect the next approximations $x^{\sigma+2}$, $x^{\sigma+3}$, ...

All these observations lead to the following modification of the algorithm. Start with an arbitrary positive vector d^0. Given the σth vector of aggregation weights d^σ and a price vector p^σ, perform the following operations:

3. Find z^σ such that

$$z^\sigma(p^\sigma, d^\sigma) = z^\sigma(p^\sigma, Ad^\sigma) + (p^\sigma, b). \tag{A3.2.3}$$

4. Find u^σ such that the vector $d^{\sigma+1} = u^\sigma(z^\sigma Ad^\sigma + b)$ has the property

$$(p^\sigma, d^{\sigma+1}) = (p^\sigma, d^\sigma). \qquad \text{(A3.2.4)}$$

Suppose that the sequence d^σ converges to a certain vector $d \neq 0$. Then $\lim_{\sigma\to\infty} z^\sigma = z$, where z is a number such that $z(p,d) = z(p,Ad) + (p,b)$. Also, $\lim_{\sigma\to\infty} u^\sigma = u$, where u is a number such that $d = u(zAd) + b$. Taking the inner product of p with both sides of the last equation, we have $(p,d) = u(z(p,Ad) + (p,b)) = uz(p,d)$. Therefore, $u = 1/z$ and $zd = A(zd) + b$. It means that $zd = x^*$ (the solution of the equation $x = Ax + b$), and $\lim_{\sigma\to\infty} x^{\sigma+1} = \lim_{\sigma\to\infty} z^\sigma Ad^\sigma + b = Azd + b = zd = x^*$. So to prove the convergence of the algorithm (A3.2.1), (A3.2.2) to x^* it suffices to prove the convergence of the algorithm (A3.2.3), (A3.2.4).

THEOREM A3.2.1 Suppose that $J = 1$, A is a productive matrix with a price vector p, p^σ is a sequence of price vectors such that $\lim_{\sigma\to\infty} p^\sigma = p$, and for every σ we have $A'p^\sigma < p^\sigma$ and d^0 is an arbitrary positive vector. Then the algorithm (A3.2.3), (A3.2.4) converges to a nonzero vector d.

Proof: First we note that if we replace all vectors $x = (x_1, \ldots, x_G)$ by $(p_1x_1, \ldots, p_Gx_G)$, all price vectors $(p_1^\sigma, \ldots, p_G^\sigma)$ by $(p_1^\sigma/p_1, \ldots, p_G^\sigma/p_G)$, and the matrix $(\hat{a}_{gq})$ by the matrix $(p_g\hat{a}_{gq}p_q^{-1})$, we will have a particular case of the theorem with $p = (1,1,\ldots,1)$. Let $x^* = (x_1^*, \ldots, x_G^*)$ be the solution of the equation $x = Ax + b$ and let $\tau = \Sigma_g x_g^*$. Replacing, if necessary, x^* by x^*/τ and the equation $x = Ax + b$ by the equation $x = Ax + b/\tau$, we can assume that $\tau = 1$.

As in the proof of Theorem A3.1.1 we have $d^{\sigma+1} = S^\sigma d^\sigma$, where $S^\sigma = (s_{gq}^\sigma)$ is a $G \times G$ matrix defined by

$$s_{gq}^\sigma = \hat{a}_{gq} + \hat{b}_g \frac{p_q^\sigma - \Sigma_\tau p_\tau^\sigma \hat{a}_{\tau q}}{\Sigma_\tau p_\tau^\sigma \hat{b}_\tau}.$$

This matrix has the following properties:

$$s_{gq}^\sigma \geq 0, \qquad (S^\sigma)'p^\sigma = p^\sigma, \qquad S^\sigma x^* = x^*.$$

We can reorder the indices 1, ..., G in such a way that the matrix S has the form

$$\begin{pmatrix} S_{11}^{\sigma} & S_{12}^{\sigma} \\ 0 & S_{22}^{\sigma} \end{pmatrix},$$

where S_{11}^{σ} is a nondecomposable $n \times n$ matrix, $n \leq G$, and S_{22}^{σ} is an $(G - n) \times (G - n)$ matrix. In addition, $(S_{11}^{\sigma})'\bar{p} = \bar{p}$ and $(S_{22}^{\sigma})'\bar{\bar{p}} < \bar{\bar{p}}$, where $\bar{p} = (p_1^{\sigma},\ldots,p_n^{\sigma})$, $\bar{\bar{p}} = (p_{n+1}^{\sigma},\ldots,p_G^{\sigma})$. This reordering will be the same for any price vector p^{σ}. If $p^{\sigma} = p$, we have a matrix

$$S = \begin{pmatrix} S_{11} & S_{12} \\ 0 & S_{22} \end{pmatrix}$$

and $S = \lim_{\sigma\to\infty} S^{\sigma}$. Again, S_{11} is a nondecomposable stochastic matrix and the ℓ^1 norm of S_{22} is less than 1 in absolute value.

As in the proof of Theorem A3.1.1, the matrix S has the following properties:

(a) $Sx^* = x^*$, where x^* is the solution of $x = Ax + b$.

(b) S leaves the hyperplane $H = \{\Sigma_g x_g = 0\}$ invariant.

Let x be an arbitrary vector. We can write $x = \tau x^* + y$, where $\tau = \Sigma_g x_g$ and $y \in H$. Then $Sx = \tau x^* + Wy$, where W is the restriction of S onto H. As in Theorem A3.1.1, all the eigenvalues of W are less than 1 in absolute value. Therefore, there exists a norm $\|\cdot\|$ on H such that $\|W\| < 1$. Note that for any price vector p^{σ}, we have $S^{\sigma}x^* = x^*$. Thus $S^{\sigma}(\tau x^* + y) = \tau x^* + (c^{\sigma},y)x^* + W^{\sigma}y$, where c^{σ} is a vector and W^{σ} is a linear transformation. Since $\lim_{\sigma\to\infty} p^{\sigma} = p$ we have $\lim_{\sigma\to\infty} c^{\sigma} = 0$ and $\lim_{\sigma\to\infty} W^{\sigma} = W$.

Consider all positive vectors x such that $\Sigma_g x_g = 1$. Write each of these vectors as $x = \tau x^* + y$ and let $M = \max \|y\|$. We can find $\beta < 1$ and $N_0 > 0$ such that for any $\sigma \geq N_0$ we have $\|c_{\sigma}\| < (1 - \beta)/2M$ and $\|W^{\sigma}\| < \beta$. Since we can start our process with p^{N_0}, d^{N_0} instead of p^0, d^0, we can assume that $N_0 = 0$.

Let

$$d^\sigma = \tau^\sigma x^* + y^\sigma = \tau^\sigma\left(x^* + \frac{y^\sigma}{\tau^\sigma}\right).$$

Since $d^\sigma > 0$, $\|y^\sigma/\tau^\sigma\| \le M$, and $\|y^\sigma\| \le M\tau^\sigma$, we have

$$d^{\sigma+1} = S^\sigma d^\sigma = [\tau^\sigma + (c^\sigma, y^\sigma)]x^* + W^\sigma y^\sigma.$$

Therefore,

$$y^{\sigma+1} = W^\sigma y^\sigma, \tag{A3.2.5}$$

$$\tau^{\sigma+1} = \tau^\sigma + (c^\sigma, y^\sigma). \tag{A3.2.6}$$

It follows from (A3.2.5) that $\|y^\sigma\| = \beta^\sigma\|y^0\|$ and $\lim_{\sigma\to\infty} \|y^\sigma\| = 0$.

On the other hand, it follows from (A3.2.6) that

$$\tau^{\sigma+1} = \tau^0 + (c^1, y^1) + (c^2, y^2) + \cdots + (c^\sigma, y^\sigma)$$

and

$$\lim_{\sigma\to\infty} \tau^\sigma = \tau^0 + \sum_{\sigma=1}^{\infty} (c^\sigma, y^\sigma).$$

The latter series is absolutely convergent because

$$\sum_\sigma |(c^\sigma, y^\sigma)| \le \frac{1-\beta}{2M}\sum_\sigma \|y^\sigma\| \le \frac{1-\beta}{2M}\|y^0\|\sum_\sigma \beta^\sigma$$
$$\le \frac{1-\beta}{2M} M\tau^0 \frac{1}{1-\beta} = \frac{\tau_0}{2}.$$

Moreover,

$$\lim_{\sigma\to\infty} \tau^{\sigma+1} > \tau^0 - \frac{\tau^0}{2} = \frac{\tau^0}{2} > 0.$$

Therefore,

$$\lim_{\sigma\to\infty} d^\sigma = (\lim_{\sigma\to\infty} \tau^\sigma)x^* \ne 0.$$

COROLLARY If $J = 1$, the algorithm (3.4.13)-(3.4.17) converges for any starting point x^0.

APPENDIX 3.3 ON THE CONVERGENCE OF THE TWO-LEVEL ITERATIVE AGGREGATION ALGORITHM THAT USES IMBALANCES MINIMIZATION CRITERION IN A SEMIAGGREGATE PROBLEM

In this section we discuss the convergence properties of the algorithms described in Section 3.5. More precisely, we discuss the convergence of a more general algorithm that proceeds as follows:

1. Perform the aggregation:

$$d_{gj}^{\sigma-1} = -\delta_{ij} x_g^{\sigma-1} + \sum_{q \in G_j} \hat{a}_{gq} x_q^{\sigma-1}, \qquad g \in G_i.$$

2. Solve the semiaggregate problem for the J-dimensional vector $z^\sigma = (z_1^\sigma, \ldots, z_J^\sigma)$; in other words, find a vector z satisfying the condition

$$\|D^{\sigma-1} z - b\| \leq \|D^{\sigma-1} e - b\|, \tag{A3.3.1}$$

where $D^{\sigma-1} = (d_{qj}^{\sigma-1})$ is a $(G \times J)$-dimensional matrix, e is the J-dimensional vector all of whose components are equal to 1, $b = (\hat{b}_g)$, $g = 1, \ldots, G$, and the norm in (A3.3.1) is an arbitrary matrix norm. To find such a vector z it suffices to perform, for example, just one step of the gradient minimization algorithm.

3. Compute the new approximation:

$$x_g^\sigma = \sum_j \sum_{q \in G_j} \hat{a}_{gq} x_q^{\sigma-1} z_j^\sigma + \hat{b}_g. \tag{A3.3.2}$$

THEOREM A3.3.1 Assume that $\|A\| < 1$, where $\|\cdot\|$ is an arbitrary matrix norm. Then the algorithm converges to a solution of the original system for any initial values x^0 and the algorithm satisfies

$$\|x^\sigma - Ax^\sigma - b\| \leq \|A\|^\sigma \|x^0 - Ax^0 - b\|.$$

Proof: Define the vector $\tilde{x}^\sigma = \{x_g^{\sigma-1} z_j^\sigma,\ g \in G_j,\ j = 1, \ldots, J\}$. We can state (A3.3.1) in terms of $\tilde{x}^\sigma$ as

$$\|\tilde{x}^\sigma - A\tilde{x}^\sigma - b\| \leq \|x^{\sigma-1} - Ax^{\sigma-1} - b\|. \tag{A3.3.3}$$

The step (A3.3.2) can be written in terms of x^σ as $x^\sigma = A\tilde{x}^\sigma + b$.

Using this expression, we have

$$\|x^\sigma - Ax^\sigma - b\| = \|A\tilde{x}^\sigma + b - A(A\tilde{x}^\sigma + b) - b\|$$
$$= \|A\tilde{x}^\sigma - A^2\tilde{x}^\sigma - Ab\| = \|A(\tilde{x}^\sigma - A\tilde{x}^\sigma - b)\|$$
$$\leqq \|A\|\|\tilde{x}^\sigma - A\tilde{x}^\sigma - b\|.$$

Combining this with (A3.3.3) yields

$$\|x^\sigma - Ax^\sigma - b\| \leqq \|A\|\|x^{\sigma-1} - Ax^{\sigma-1} - b\|.$$

The proof of the theorem follows from this inequality. The convergence rates discussed in Section 3.5 follow by using the ℓ_1 or ℓ_2 norms in the theorem.

APPENDIX 3.4 MATHEMATICAL ANALYSIS OF THE GENERALIZATIONS, VARIANTS, AND MODIFICATIONS OF THE BASIC ITERATIVE AGGREGATION ALGORITHMS FOR INTERPRODUCT INPUT-OUTPUT MODELS

We present some of the sufficient conditions for the convergence of the algorithms described in Section 3.6.

THEOREM A3.4.1 The generalized two-level iterative aggregation algorithm (3.6.1)-(3.6.3), (3.6.5) converges if

$$\|A\| + 2\|D\| < 1.$$

Proof: Let x denote the solution of the original system (3.1.1), $\alpha = \|A\|$ and $\delta = \|D\|$, and set $x^\sigma = x + \varepsilon^\sigma$, $\tilde{x}^\sigma = x + \tilde{\varepsilon}^\sigma$, $z^\sigma = 1 + t^\sigma$. Then (3.6.2) and (3.6.5) can be written as

$$t_i^\sigma X_i^{\sigma-1} = \sum_{g\in G_i} - \varepsilon_g^{\sigma-1} + \sum_{g\in G_i} \sum_q \hat{a}_{gq}\varepsilon_q^{\sigma-1}$$
$$+ \sum_j t_j^\sigma \sum_{g\in G_i} \sum_{q\in G_j} \hat{a}_{gq}x_q^{\sigma-1}; \tag{A3.4.1}$$

$$\varepsilon_g^\sigma = \Sigma_q d_{gq}\tilde{\varepsilon}_q^\sigma. \tag{A3.4.2}$$

Let us find an estimate for $\|\tilde{\varepsilon}^\sigma\|$. Simple algebra yields the inequality

$$\|\tilde{\varepsilon}^\sigma\| \leqq \Sigma_i |t_i^\sigma| X_i^{\sigma-1} + \|\varepsilon^{\sigma-1}\|. \tag{A3.4.3}$$

Indeed,

$$\|\tilde{\varepsilon}^{\sigma}\| = \|\tilde{x}_g^{\sigma} - x_g\| = \sum_i \sum_{g \in G_i} \|z_i^{\sigma} x_g^{\sigma-1} - x_g\|$$

$$= \sum_i \sum_{g \in G_i} |t_i^{\sigma} x_g^{\sigma-1} + \varepsilon_g^{\sigma-1}| \leqq \sum_i |t_i^{\sigma}| X_i^{\sigma-1} + \|\varepsilon^{\sigma-1}\|.$$

From (A3.4.1) we obtain

$$|t_i^{\sigma}| X_i^{\sigma-1} \leqq \sum_{g \in G_i} |\varepsilon_g^{\sigma-1}| + \sum_{g \in G_i} \sum_q \hat{a}_{gq} |\varepsilon^{\sigma-1}| + \sum_j |t_j^{\sigma}| \sum_{g \in G_i} \sum_{q \in G_j} \hat{a}_{gq} x_q^{\sigma-1}. \quad \text{(A3.4.4)}$$

Summing (A3.4.4) over i (i = 1, ..., J), we have

$$\Sigma_i \; |t_i^{\sigma}| X_i^{\sigma-1} \leqq (1 + \alpha)\|\varepsilon^{\sigma-1}\| + \alpha \, \Sigma_j \; |t_j^{\sigma}| X_j^{\sigma-1}$$

or

$$\sum_i |t_i^{\sigma}| X_i^{\sigma-1} \leqq \frac{(1 + \alpha)\|\varepsilon^{\sigma-1}\|}{1 - \alpha}. \quad \text{(A3.4.5)}$$

Substituting (A3.4.5) into (A3.4.3), we obtain the bound

$$\|\tilde{\varepsilon}^{\sigma}\| \leqq \frac{2\|\varepsilon^{\sigma-1}\|}{1 - \alpha}. \quad \text{(A3.4.6)}$$

From (A3.4.2) it follows that

$$\|\varepsilon^{\sigma}\| \leqq \delta \|\tilde{\varepsilon}^{\sigma}\|. \quad \text{(A3.4.7)}$$

Combining (A3.4.6) and (A3.4.7), we obtain the final bound:

$$\|\varepsilon^{\sigma}\| \leqq \frac{2\delta}{1 - \alpha} \|\varepsilon^{\sigma-1}\|.$$

Thus convergence is assured if $2\delta/(1 - \alpha) < 1$ or $\alpha + 2\delta < 1$, which completes the proof of the theorem.

The sufficient condition for the convergence of the basic two-level iterative aggregation algorithm in Appendix 3.1 follows from this theorem, as do sufficient conditions for the variants of the two-level algorithm when the norm of the matrix is sufficiently small, as the matrix D is continuously dependent on the matrix A and if α is sufficiently small, the condition of the theorem clearly is valid.

THEOREM A3.4.2 The generalized one-level process (3.3.5), (3.6.13), (3.6.14) converges if $\alpha + \delta + \hat{\delta} + \alpha(\hat{\delta} - \delta) < 1$, where $\hat{\delta} = \Sigma_i \max_q \Sigma_{g\in G_i} d_{gq}$.

Proof: Using the notation of Theorem A3.4.1, the original process can be restated as

$$t_i^{s,\sigma}X_i^{\sigma-1} = -\sum_{g\in G_i} \varepsilon_g^{\sigma-1} + \sum_{g\in G_i}\sum_q \hat{a}_{gq}\varepsilon_q^{\sigma-1} + \sum_{j\neq s} t_j^{s,\sigma} \sum_{g\in G_i}\sum_{q\in G_j} \hat{a}_{gq}x_q^{\sigma-1}, \tag{A3.4.8}$$

$$\varepsilon_g^{\sigma} = \sum_q d_{gq}\varepsilon_q^{\sigma-1} + \sum_{j\neq s} t_j^{s,\sigma} \sum_{q\in G_j} d_{gq}x_q^{\sigma-1}. \tag{A3.4.9}$$

From (A3.4.8) it follows that

$$\sum_{i\neq s} |t_i^{s,\sigma}|X_i^{\sigma-1} \leqq (1 + \alpha)\|\varepsilon^{\sigma-1}\| + \alpha \sum_{j\neq s} |t_j^{s,\sigma}|X_j^{\sigma-1}.$$

This implies that

$$|t_i^{s,\sigma}|X_i^{\sigma-1} \leqq \frac{(1 + \alpha)\|\varepsilon^{\sigma-1}\|}{1 - \alpha} \quad \text{for all } s. \tag{A3.4.10}$$

Using (A3.4.9), we obtain

$$\sum_{g\in G_s} |\varepsilon_g^{\sigma}| \leqq \sum_{g\in G_s}\sum_q d_{gq}|\varepsilon_q^{\sigma-1}| + \delta_s \sum_{i\neq s} |t_i^{s,\sigma}|X_i^{\sigma-1}, \tag{A3.4.11}$$

where $\delta_s = \max_q \Sigma_{g\in G_s} d_{gq}$. Summing (A3.4.11) over s and combining this with (A3.4.10), we obtain

$$\|\varepsilon^{\sigma}\| \leqq \|\varepsilon^{\sigma-1}\|\left[\delta + \frac{(1 + \alpha)\hat{\delta}}{1 - \alpha}\right].$$

Thus convergence is guaranteed as long as

$$\delta + \frac{(1 + \alpha)\hat{\delta}}{1 - \alpha} < 1,$$

which is equivalent to the condition of the theorem.

THEOREM A3.4.3 The generalized algorithm (3.3.5), (3.6.12) converges if

$$\alpha + 2\delta + \hat{\delta} + 2\alpha(\hat{\delta} - \delta) < 1.$$

Proof: Similarly to the equalities (A3.4.8) and (A3.4.9) of Theorem A3.4.2, we derive

$$t_i^{s,\sigma} X_i^{\sigma-1} = -\sum_{g \in G_i} \varepsilon_g^{\sigma-1} + \sum_{g \in G_i} \sum_{q \in G_s} \hat{a}_{gq} \varepsilon_q^{\sigma} + \sum_{g \in G_i} \sum_{q \notin G_s} \hat{a}_{gq} \varepsilon_q^{\sigma-1}$$

$$+ \sum_{j \neq s} t_j^{s,\sigma} \sum_{g \in G_i} \sum_{q \in G_j} \hat{a}_{gq} x_q^{\sigma-1}, \qquad i \neq s; \qquad \text{(A3.4.12)}$$

$$\varepsilon_g^{\sigma} = \sum_{q \in G_s} d_{gq} \varepsilon_q^{\sigma} + \sum_{q \notin G_s} d_{gq} \varepsilon_q^{\sigma-1} + \sum_{j \neq s} t_j^{s,\sigma} \sum_{q \in G_j} d_{gq} x_q^{\sigma-1},$$

$$g \in G_s. \qquad \text{(A3.4.13)}$$

Again as in Theorem A3.4.2, from (A3.4.12), (A3.4.13) we have

$$\|\varepsilon^{\sigma}\| \leq \left[1 - \delta - \frac{\hat{\delta}(1 + \alpha)}{1 - \alpha}\right]^{-1} \left(\delta + \frac{\alpha\hat{\delta}}{1 - \alpha}\right) \|\varepsilon^{\sigma-1}\|.$$

Convergence is ensured if the coefficient at the right-hand side of this equation is less than 1. This is the condition of the theorem.

THEOREM A3.4.4 The modified algorithm (3.6.1), (3.6.15) converges[1] if $3\alpha - \alpha^{\tau+1} - \alpha^{\tau+2} < 1$.

Proof: Using the same notation as Theorems A3.4.2 and A3.4.3, the original process can be restated as

$$t_i^{\tau,\sigma} = (X_i^{\sigma-1})^{-1} \left[\sum_j X_{ij}^{\sigma-1} t_j^{\tau-1,\sigma} + \sum_{g \in G_i} \sum_q \hat{a}_{gq} \varepsilon_q^{\sigma-1} - \sum_{q \in G_i} \varepsilon_q^{\sigma-1}\right],$$

(A3.4.14)

---

[1]Remember that $\alpha < 1$ because of the productivity conditions.

$$\varepsilon_g^{\sigma} = \sum_q \hat{a}_{gq}\varepsilon_q^{\sigma-1} + \sum_j t_j^{\tau,\sigma} \sum_{q\in G_j} \hat{a}_{gq}x_q^{\sigma-1}. \tag{A3.4.15}$$

We introduce the following matrices:

$$V = (v_{iq}), \qquad v_{iq} = \begin{cases} 1 & \text{if } q \in G_i; \\ 0 & \text{if } q \notin G_i. \end{cases}$$

$$X^{\sigma-1} = \{x_{gj}^{\sigma-1}\}, \qquad x_{gj}^{\sigma-1} = \begin{cases} x_g^{\sigma-1} & \text{if } g \in G_j; \\ 0 & \text{if } g \notin G_j. \end{cases}$$

$$D^{\sigma-1} = \{d_{ij}^{\sigma-1}\}, \qquad d_{ij}^{\sigma-1} = \begin{cases} X_i^{\sigma-1} & \text{if } i = j; \\ 0 & \text{if } i \neq j. \end{cases}$$

The matrix $\tilde{A}^{\sigma-1}$ is defined as $\tilde{A}^{\sigma-1} = VAX^{\sigma-1}(D^{\sigma})^{-1}$. Then in this notation (A3.4.14), (A3.4.15) can be rewritten as

$$t^{\tau,\sigma} = (D^{\sigma-1})^{-1}\tilde{A}^{\sigma-1}D^{\sigma-1}t^{\tau-1,\sigma} + (D^{\sigma-1})^{-1}V(A - I)\varepsilon^{\sigma-1}. \tag{A3.4.16}$$

$$t^{0,\sigma} = 0.$$

$$\varepsilon^{\sigma} = A\varepsilon^{\sigma-1} + AX^{\sigma-1}t^{\tau,\sigma}. \tag{A3.4.17}$$

From (A3.4.16) we have

$$t^{\tau,\sigma} = (D^{\sigma-1})^{-1}[(\tilde{A}^{\sigma-1})^{\tau-1} + (\tilde{A}^{\sigma-1})^{\tau-2} + \cdots + I]V(A - I)\varepsilon^{\sigma-1}. \tag{A3.4.18}$$

Substituting this value of $t^{\tau,\sigma}$ into (A3.4.17) yields

$$\varepsilon^{\sigma} = \{A + AX^{\sigma-1}(D^{\sigma-1})^{-1}[(\tilde{A}^{\sigma-1})^{\tau-1} + \cdots + I]V(A - I)\}\varepsilon^{\sigma-1}. \tag{A3.4.19}$$

A sufficient condition for convergence of the process is that the norm β of the matrix in brackets in equation (A3.4.19) is less than 1. It is easy to verify that $\|X^{\sigma-1}(D^{\sigma-1})^{-1}\| \leqq 1$ and that $\|V\| = 1$. Then β will have the following upper bound:

$$\beta \leqq \alpha + \alpha(\alpha^{\tau-1} + \alpha^{\tau-2} + \cdots + 1)(\alpha + 1)$$

$$= \frac{\alpha}{1 - \alpha}(2 - \alpha^{\tau} - \alpha^{\tau+1}).$$

The condition of the theorem is equivalent to this last term having a value less than 1.

THEOREM A3.4.5 The modified two-level process (3.6.1), (3.6.16), (3.6.17) converges if $\alpha < 1/3$ and k is contained in the interval

$$0 \leqq k < \frac{1 - 3\alpha}{\alpha + \alpha^2}. \tag{A3.4.20}$$

Proof: Inequality (3.6.16) can be restated as

$$X_i^{\sigma-1} z_i^\sigma = \Sigma_j X_{ij}^{\sigma-1} z_j^\sigma + \tilde{b}_i + \delta_i^\sigma, \tag{A3.4.21}$$

$$\|\delta^\sigma\| < k\|x^{\sigma-1} - Ax^{\sigma-1} - b\|. \tag{A3.4.22}$$

In terms of the variables t, ε and the matrices $A^{\sigma-1}$, $D^{\sigma-1}$, V of Theorem A3.4.4, we have from (A3.4.22)

$$t^\sigma = (D^{\sigma-1})^{-1}\tilde{A}^{\sigma-1}D^{\sigma-1}t^\sigma + (D^{\sigma-1})^{-1}[V(A - I)\varepsilon^{\sigma-1} + \delta^\sigma],$$

from which it follows that

$$t^\sigma = (D^{\sigma-1})^{-1}(I - \tilde{A}^{\sigma-1})^{-1}[V(A - I)\varepsilon^{\sigma-1} + \delta^\sigma]. \tag{A3.4.23}$$

Substituting this last value into (A3.4.17), we have

$$\varepsilon^\sigma = A\varepsilon^{\sigma-1} + AX^{\sigma-1}(D^\sigma)^{-1}(I - \tilde{A}^{\sigma-1})[V(A - I)\varepsilon^{\sigma-1} + \delta^\sigma]. \tag{A3.4.24}$$

Clearly, $\|\tilde{A}^{\sigma-1}\| \leqq \alpha$, and therefore $\|I - \tilde{A}^{\sigma-1}\|^{-1} \leqq 1/(1 - \alpha)$. From this we obtain

$$\|\varepsilon^\sigma\| \leqq \alpha\|\varepsilon^{\sigma-1}\| + \alpha(1 - \alpha)^{-1}[(1 + \alpha)\|\varepsilon^{\sigma-1}\| + \|\delta^\sigma\|]. \tag{A3.4.25}$$

Condition (A3.4.22) implies that

$$\|\delta^\sigma\| \leqq k\|\varepsilon^{\sigma-1} - A\varepsilon^{\sigma-1}\| \leqq k(1 + \alpha)\|\varepsilon^{\sigma-1}\|. \tag{A3.4.26}$$

Combining this with (A3.4.25), we obtain the inequality

$$\|\varepsilon^\sigma\| \leqq [\alpha + \alpha(1 - \alpha)^{-1}(1 + \alpha)(1 + k)]\|\varepsilon^{\sigma-1}\|. \tag{A3.4.27}$$

Thus the process will converge if

$$\alpha + \alpha(1 - \alpha)^{-1}(1 + \alpha)(1 + k) < 1. \qquad \text{(A3.4.28)}$$

Simple algebra shows that (A3.4.28) is equivalent to the condition (A3.4.20) which proves the theorem.

4

Iterative Aggregation Algorithms for Solving General Linear Systems

4.1 TWO-LEVEL ITERATIVE AGGREGATION ALGORITHMS WITH MINIMIZATION OF IMBALANCES IN A SEMIAGGREGATE PROBLEM

Assume that we are given a general linear system

$$Ax = b. \tag{4.1.1}$$

When the coefficient matrix A is a given matrix, then even if A and the vector b are restricted to be positive, the solution of the system of equations may have negative components. Despite this, in our analysis of the general algorithms we will continue to call the components of the vector b resources and the elements $\hat{a}_{gq}$ of the matrix A the input coefficients per unit of output for each type of product. Thus the problem (4.1.1) will still be interpreted as finding the plan x that guarantees the full utilization of the resource vector b.

We noted in Chapter 3 that iterative aggregation algorithms for solving the problem of balancing interindustry proportions can proceed by decreasing at each iteration the general imbalance of the plan obtained at the previous iteration. A similar formulation can be made for the industry problem. This is accomplished by having industry j attempt to improve the balance of the constraint set by assuming that all other industries have taken the new solution by formula (3.2.11). As the norm of the imbalance vector reflects the degree of imbalance in the present solution, industry j wants to minimize this norm with respect to x_j, that is, to solve

$$\|A_j x_j - (b - \Sigma_{i \neq j} A_i \tilde{x}_i^{\sigma})\| \to \min, \tag{4.1.2}$$

where A_i is the submatrix of A consisting of the columns of A with indices in G_i and where $\tilde{x}_i^\sigma$ is defined as $\tilde{x}_i^\sigma = (\tilde{x}_g^\sigma | g \in G_i)$, $\tilde{x}_g^\sigma = x_g^{\sigma-1} z_i^\sigma$.

When the Euclidean norm is used, one iteration of the algorithm consists of the following steps:

1. Compute the semiaggregate product flows:

$$x_{gj}^{\sigma-1} = \sum_{q \in G_j} \hat{a}_{gq} x_q^{\sigma-1}. \tag{4.1.3a}$$

2. Solve the minimization problem for imbalances:

$$\Sigma_g (\Sigma_j z_j x_{gj}^{\sigma-1} - \hat{b}_g)^2 \to \min. \tag{4.1.3b}$$

3. Find new x_g^σ by minimizing the imbalances for each industry j:

$$\min \Delta^2 = \min \sum_g \left(\sum_{q \in G_j} \hat{a}_{gq} x_q - \hat{b}_g + \sum_{i \neq j} x_{gi}^{\sigma-1} z_i^\sigma \right)^2. \tag{4.1.3c}$$

As in Section 3.5, we can show that (4.1.3c) is equivalent to solving the system

$$\sum_g \hat{a}_{gq} \left(\sum_{k \in G_j} \hat{a}_{gk} x_k - \hat{b}_g + \sum_{i \neq j} x_{gi}^{\sigma-1} z_i^\sigma \right) = 0, \qquad q \in G_j.$$

In matrix form this can be written as

$$A_j' A_j x_j = A_j' \left(b - \sum_{i \neq j} A_i x_i^{\sigma-1} z_i^\sigma \right). \tag{4.1.4}$$

Note that the matrix $A_j' A_j$ can be computed in advance and be kept by industry j. The vector $b_j' = b - \Sigma_{i \neq j} A_i x_i^{\sigma-1} z_i^\sigma$ can be calculated as follows. Industry i, using the parameters $x_{gi}^{\sigma-1}$, $g \in G_i$ and z_i^σ, $i = 1, \ldots, J$, computes $b_g'' = \hat{b}_g - \Sigma_r x_{gr}^{\sigma-1} z_r^\sigma$, $g \in G_i$, and passes these values to the other industries. Having received b_g'', industry j then determines the components b_{gj}' of the vector b_j' by the formula

$$\hat{b}_{gj}' = b_g'' + x_{gj}^{\sigma-1} z_j^\sigma$$

and then multiplies b_j' by the matrix A_j'. In comparison with the algorithm in Section 3.5, this requires the industries to pass at each iteration an additional (J - 1)G numbers while the amount of information passed between the industries and the center remain unchanged.

We can modify this algorithm so that the industries' decisions reflect to a greater degree the recommendations of the center at the present iteration. This is done by having industry j calculate x_g^σ by averaging over the values generated by solving (4.1.2) and the values recommended by the center, that is,

$$x_g^\sigma = \gamma_j^\sigma \hat{x}_g^\sigma + (1 - \gamma_j^\sigma) x_g^{\sigma-1} z_j^\sigma, \qquad g \in G_j, \tag{4.1.5}$$

Here $\hat{x}_g^\sigma$ is the solution of the minimization problem (4.1.4) and $\gamma_j^\sigma > \delta > 0$, $\Sigma_j \gamma_j^\sigma = 1$. It is proven in Appendix 4.1 that this modified process with averaged disaggregate values converges globally.

In the algorithms discussed so far all the industries determine their new approximations to the output vectors simultaneously. It is easy to construct a process in which the industries compute their new approximate output vectors sequentially. This would be done by replacing (4.1.4) with

$$A_j' A_j x_j = A_j' \left(b - \sum_{i<j} A_i x_i^\sigma - \sum_{i>j} A_i \tilde{x}_i^\sigma \right). \tag{4.1.6}$$

Although this algorithm appears to have a faster rate of convergence than the previous algorithm, it is less convenient to use in practice, as the calculations for each industry cannot be performed in parallel. This algorithm converges for any given matrix A and vector b in the original problem.

Mathematically, these algorithms can be seen to be incorporating iterative aggregation into group coordinate descent methods. Similarly to the way that iterative aggregation was built in as a correcting step into different algorithms for solving Ax + b = x, for an aggregated problem we can insert a correction into many algorithms for solving (4.1.1) that use minimization of imbalances. So in addition to the algorithms described in this section, there are interesting algorithms that use one step of the steepest descent method to calculate x^σ from $\tilde{x}^\sigma$.

4.2 ONE-LEVEL ITERATIVE AGGREGATION ALGORITHMS WITH MINIMIZATION OF IMBALANCES IN A SEMIAGGREGATE PROBLEM

In both the original two-level algorithm (3.2.9)-(3.2.11) as well as in one-level algorithm (3.3.5), (3.3.6), the aggregate problems are formed so that the total balance is always satisfied. It is also natural to try to determine x_g, $g \in G_i$, and z_r^{i}, for industry i such that the original equations are as close as possible to being in balance. This leads to an algorithm based on using the Euclidean norm. Industry j will then try to solve the problem

$$\min_{x_q, z_i^j} \sum_g \left(\sum_{q \in G_j} \hat{a}_{gq} x_q + \sum_{i \neq j} x_{gi}^{\sigma-1} z_i^j - \hat{b}_g \right)^2. \tag{4.2.1}$$

Denote the solution of (4.2.1) by $\hat{x}_g^\sigma$, $g \in G_j$, and $z_r^{j,\sigma}$. For the new approximate solution each industry calculates the weighted mean of its own solution $\hat{x}_g^\sigma$ and the demands of the other industries:

$$\begin{aligned} x_g^\sigma &= \gamma_j^\sigma \hat{x}_g^\sigma + \sum_{i \neq j} \gamma_i^\sigma z_j^{i,\sigma} x_g^{\sigma-1}, \qquad g \in G_j; \\ \Sigma_i \gamma_i^\sigma &= 1, \qquad \gamma_i^\sigma > \delta > 0. \end{aligned} \tag{4.2.2}$$

This algorithm converges to the solution of the original problem for any given A and b. To study this algorithm in more detail, it is convenient to rewrite it in matrix form. As before, we let A_j denote the submatrix of A made up of the columns of A whose indices are in G_j. In this notation, (4.2.1) becomes

$$\min \left\| A_j x_j + \sum_{r \neq j} (A_r x_r^{\sigma-1}) z_r^j - b \right\|. \tag{4.2.3}$$

Here $x_j^{\sigma-1} = \{x_g^{\sigma-1}, g \in G_j\}$. Note that $A_j x_j^{\sigma-1} = \{x_{gj}^{\sigma-1}, g = 1, \ldots, G\}$. Let $\hat{A}_j^{\sigma-1}$ denote the matrix whose columns are formed from the vectors $A_r x_r^{\sigma-1}$, $r \neq j$. Then (4.2.3) can be written more simply as

$$\min \|A_{\cdot j}x + \hat{A}_{\cdot j}^{\sigma-1} z^j - b\|, \tag{4.2.4}$$

where $z^j = \{z_r^j, r \neq j\}$. Since the norm in (4.2.4) has been defined to be the Euclidean norm, the solution to (4.2.4) can be found by solving the following system of equations (as was shown above for the two-level algorithm):

$$\begin{aligned} &\tilde{A}_{11}^{j} x_j + \tilde{A}_{12}^{j,\sigma-1} z^j = \bar{b}_1; \\ &(\tilde{A}_{12}^{j,\sigma-1})' x_j + \tilde{A}_{22}^{j,\sigma-1} z^j = \bar{b}_2^{\sigma-1}. \end{aligned} \tag{4.2.5}$$

Here $\tilde{A}_{11}^j = A_{\cdot j}' A_{\cdot j}$, $\tilde{A}_{12}^{j,\sigma-1} = A_{\cdot j}' \hat{A}_{\cdot j}^{\sigma-1}$, $\tilde{A}_{22}^{j,\sigma-1} = (\hat{A}_{\cdot j}^{\sigma-1})' \hat{A}_{\cdot j}^{\sigma-1}$, $\bar{b}_1 = A_{\cdot j}' b$, $\bar{b}_2^{\sigma-1} = (\hat{A}^{\sigma-1})' b$. Note that $\tilde{A}_{11}^j$ and $\bar{b}_1$ can be computed in advance, while the terms $\tilde{A}_{22}^{j,\sigma-1}$, $\tilde{A}_{12}^{j,\sigma-1}$, and $\bar{b}_2^{\sigma-1}$ can be computed in two stages as in algorithm (3.3.5), (3.3.6). The first stage involves computation of parts of the matrix by each industry; the second stage involves the passing of these results to industry j, which then forms (4.2.5).

This algorithm allows parallel computations. It is clear that it is also possible to construct a sequential version of this algorithm in which x_g, $g \in G_j$, will be derived from the following:

$$\min \left\|A_{\cdot j}x_j + \sum_{\tau<j} (A_r x_r^{\sigma}) z_r^j + \sum_{r>j} (A_r x_r^{\sigma-1}) z_r^j - b\right\|. \tag{4.2.6}$$

4.3 TWO-LEVEL ALGORITHMS WITH MINIMIZATION OF IMBALANCES AND AGGREGATION OF INCREMENTS

In this section we analyze iterative aggregation algorithms for solving (4.1.1) that are based on aggregating incremental changes to the variables rather than the variables themselves as in the previous sections. These incremental algorithms can be used to solve problems somewhat more general than (4.1.1). However, initially we restrict our discussion to solving (4.1.1), owing to the importance of this problem and the fact that it makes each step of the algorithm analytically explicit.

The algorithms discussed in this section are generalizations of several classical algorithms for solving (4.1.1) in which one-dimensional minimization is used at each iteration. Analogously to Sections 4.1 and 4.2, these algorithms solve (4.1.1) by minimizing the function

$$F(x) = (Ax - b)^2,$$
$$A = \{\hat{a}_{gq} | g, q = 1, \ldots, G\}, \qquad b = (\hat{b}_g | g = 1, \ldots, G). \tag{4.3.1}$$

As before, we assume that $\{1,\ldots,G\} = \cup_{j=1}^{J} G_j$, $\bar{b}_j = (\hat{b}_g | g \in G_j)$, $A_j = (\hat{a}_{gq} | g = 1, \ldots, G, q \in G_j)$, $A_{ij} = (\hat{a}_{gq}, g \in G_i, q \in G_j)$. At the completion of iteration $\sigma - 1$ we have the approximate solution $x^{\sigma-1} = (x_q^{\sigma-1} | q = 1, \ldots, G)$; that is, each subsystem j has an approximate solution to its variables $x_j^{\sigma-1} = (x_q^{\sigma-1} | q \in G_j)$. The basic outline of the new algorithm is as follows. First, each subsystem determines a desirable VDI. The following example illustrates this point. Suppose that the first subsystem contains the variables x_1, x_2, x_3, and $x_1^{\sigma-1} = 3$, $x_2^{\sigma-1} = 0.5$, and $x_3^{\sigma-1} = 20$. Then if the subsystem chooses as its VDI the vector (4,4,1), this implies that the subsystem desires to change x_1 and x_2 by the same amount, whereas x_3 will be incremented by a factor one-fourth that of the other two components.

Second, after determining a desirable VDI $\Delta x^{\sigma-1} = (\Delta x_q^{\sigma-1} | q \in G_j$, $j = 1, \ldots, J)$ it is necessary to find the increments that yield the greatest increase in the objective function. Thus the main difference between this algorithm and those discussed previously is that at iteration $\sigma - 1$ we must calculate the vector $x^{\sigma-1}$ as well as the vector $\Delta x^{\sigma-1} = (\Delta x_g^{\sigma-1} | g = 1, \ldots, G)$.

A more formal description of the algorithm is:

1. Perform the aggregation. Subsystem j computes its resource needs for its present approximation to the plan

$$b_{gj}^{\sigma-1} = \sum_{q \in G_j} \hat{a}_{gq} x_q^{\sigma-1}, \qquad g = 1, \ldots, G; \tag{4.3.2}$$

then it computes the semiaggregate inputs of resource g for the increment $\Delta x_j^{\sigma-1}$ of subsystem j's output:

$$\tilde{a}_{gj}^{\sigma-1} = \sum_{q \in G_j} \hat{a}_{gq} \Delta x_q^{\sigma-1}, \qquad g = 1, \ldots, G; \qquad (4.3.3)$$

computes the vector of unused resources:

$$\tilde{b}_g^{\sigma-1} = \hat{b}_g - \Sigma_i \, b_{gi}^{\sigma-1}; \qquad (4.3.4)$$

computes the contribution of subsystem r to the aggregated coefficient matrix $D^{\sigma-1}$:

$$d_{ij}^{r,\sigma-1} = \sum_{g \in G_r} \tilde{a}_{gi}^{\sigma-1} \tilde{a}_{gj}^{\sigma-1}, \qquad i, r = 1, \ldots, J; \qquad (4.3.5)$$

and computes the contribution of subsystem r to the aggregate constant vector:

$$b_i^{r,\sigma-1} = \sum_{g \in G_r} \tilde{b}_g^{\sigma-1} \tilde{a}_{gi}^{\sigma-1}, \qquad i = 1, \ldots, J. \qquad (4.3.6)$$

2. Formulate the aggregate problem. The center adds up the incoming information and determines

$$d_{ij}^{\sigma-1} = \Sigma_r \, d_{ij}^{r,\sigma-1}, \qquad i, j = 1, \ldots, J \qquad (4.3.7)$$

and

$$b_i^{\sigma-1} = \Sigma_r \, b_i^{r,\sigma-1}, \qquad i = 1, \ldots, J. \qquad (4.3.8)$$

3. Solve the aggregate problem:

$$\Sigma_j \, d_{ij}^{\sigma-1} z_j = b_i^{\sigma-1}. \qquad (4.3.9)$$

Denote the solution by the vector $z^\sigma = (z_j^\sigma | j = 1, \ldots, J)$.

4. Compute the new approximation:

$$x_g^\sigma = x_g^{\sigma-1} + z_j^\sigma \Delta x_g^{\sigma-1}, \qquad g \in G_j. \qquad (4.3.10)$$

5. Determine a new VDI.

There are several methods for determining the new VDI and the choice of Δx^σ determines the type of algorithm. For example, setting Δx^σ equal to the vector $A'(b - Ax^\sigma) = A'\tilde{b}^\sigma$ sets the direction collinear with the gradient of $F(x)$.

Another way to choose Δx_j^σ is minimize $(Ax - b)^2$ assuming that the increments to the variables of the other subsystems are zero. Then Δx_j^σ is determined by solving

$$A_j'A_j\Delta x_j^\sigma = A_j'\tilde{b}^\sigma. \tag{4.3.11}$$

The matrix $A_j'A_j$ can be computed in advance and the vector $A_j'\tilde{b}^\sigma$ can be computed by a method similar to the analog of the gradient method discussed previously. Thus, calculating Δx_j^σ from (4.3.11) is computationally about the same as choosing the appropriate component of the gradient vector.

4.4 ONE-LEVEL ALGORITHMS WITH MINIMIZATION OF IMBALANCES AND AGGREGATION OF INCREMENTS

In the parallel one-level algorithm, at iteration σ the subsystems solve the following:

1. Perform the aggregation. Subsystem j computes the upper-level inputs $A_jx_j^{\sigma-1}$ and $A_j\Delta x_j^{\sigma-1}$ and passes the vectors $A_{ij}\Delta x_j^{\sigma-1}$, $A_{ij}x_j^{\sigma-1}$ to subsystem i. After all subsystems have passed this information, subsystem j calculates $\tilde{b}_j^{\sigma-1} = \bar{\bar{b}}_j - \Sigma_i A_{ji}x_i^{\sigma-1}$ and these values are passed to each subsystem. Subsystem j then forms the matrix $\hat{A}_j^{\sigma-1}$ whose columns consist of the vectors $A_ix_i^{\sigma-1}$, $i \neq j$.
2. Determine the aggregate system:

$$\begin{aligned} A_j'A_jx_j + A_j'\hat{A}_j^{\sigma-1}z &= A_j'\tilde{b}_j^{\sigma-1}; \\ (\hat{A}_j^{\sigma-1})'A_jx_j + (\hat{A}_j^{\sigma-1})'\hat{A}_j^{\sigma-1}z &= \hat{A}_j^{\sigma-1}\tilde{b}_j^{\sigma-1}. \end{aligned} \tag{4.4.1}$$

Note that $(\hat{A}_j^{\sigma-1})'A_j = (A_j'\hat{A}_j^{\sigma-1})'$, while $A_j'A_j$ and $(\hat{A}_j^{\sigma-1})'\hat{A}_j^{\sigma-1}$ are symmetric matrices. Denote the solution of (4.4.1) by $\tilde{x}_j^\sigma$, $z_i^{j,\sigma}$,

$i \neq j$. The solution $z_i^{j,\sigma}$ is passed to subsystem i.

3. Compute the new approximation:

$$x_g^\sigma = \gamma_j^\sigma \tilde{x}_g^\sigma + \sum_{i \neq j} \gamma_i (x_g^{\sigma-1} + z_j^{i,\sigma} \Delta x_g^{\sigma-1}), \qquad g \in G_j, \tag{4.4.2}$$

where $\gamma_j^\sigma > \delta > 0$, $\Sigma_j \gamma_j^\sigma = 1$ for all σ. A typical value for γ_j^σ is $1/J$.

In the sequential algorithm the subsystems determine this approximation in turn. In this case subsystem j obtains $A_i x_i^\sigma$, $A_i \Delta x_i^\sigma$, $i < j$, and $A_i x_i^{\sigma-1}$, $A_i \Delta x_i^{\sigma-1}$, $i > j$, and takes as x_j^σ the solution of (4.4.1) with the appropriate changes in $\hat{A}_j^{\sigma-1}$, $\tilde{b}_j^{\sigma-1}$, without performing step 3. The vectors Δx_i^σ are determined in the one-level algorithm in the same way as in the two-level algorithm.

The parallel algorithms based on group coordinate descent methods underlie our related algorithms for solving the general unconstrained optimization problem. For such problems, the group coordinate descent algorithms oftem prove as effective as the other algorithms, particularly when the present estimate is far from the actual solution, and often are quite useful in the case where J is small.

4.5 NUMERICAL EXPERIMENTS WITH ITERATIVE AGGREGATION ALGORITHMS FOR LINEAR SYSTEMS OF EQUATIONS

The primary interest in iterative aggregation methods lies in the fact that they make it possible to coordinate the solution of problems containing differing levels of aggregation. Still it is of interest to compare the computational efficiency of iterative aggregation algorithms with other iterative methods for solving similar problems. The computational criteria we have in mind include the number of iterations required by each method to achieve a given degree of accuracy in the solution vector from a generic starting point, the number of operations per iteration and the total number of operations required to obtain a solution with a given degree of accuracy, and the total time it takes to obtain a solution with a given degree of accuracy with due regard for the computational time needed to pass the necessary data arrays between iterations.

Although the last criterion is in some sense a general one, the other criteria may have a significance of their own. For example, at a certain level of computer technology and communications, the number of iterations may be the most important factor. In this section we present the results from a number of numerical experiments conducted on computers using the one- and two-level iterative aggregation algorithms described in Chapters 3 and 4 for cost interproduct balances. These experiments were performed to get some idea of the actual convergence rates of these algorithms in terms of the number of iterations.

A great number of numerical experiments were performed to compare the basic two-level iterative aggregation process with the method of successive approximations and with Gauss-Seidel iterations. These results are reported in Arkhangelsky et al. (1975). The numerical experiments show that the basic one-level iterative aggregation algorithm requires 1.5 to 2.0 times fewer iterations than the basic two-level algorithm for the same degree of accuracy and that the two-level algorithm converges 1.5 to 2.0 times faster than does successive approximations. The advantages over Gauss-Seidel iterations are somewhat less impressive, however, as the basic iterative aggregation algorithms incorporate successive approximations and are designed for subsystems operating in parallel, the comparison with Gauss-Seidel iterations is not as relevant.

The numerical experiments also have shown a weak correlation between the rate of convergence and the number of industries J when the value of J is less than half the number of commodities G, provided that each industry contains the same number of commodities. These experiments have been conducted with coefficient matrices of dimensions ranging from 18 to 100 and for arbitrary matrices that have a norm less than 1 (experiments conducted on matrices that have a norm of 1 have produced similar results).

At each iteration the two-level iterative aggregation algorithm requires more operations than does the successive approximations algorithm, and the one-level iterative aggregation algorithm requires more operations than either of the other algorithms. However, given the present state of development in both the field of communications and in parallel computations, the number of iterations required to find an acceptable solution would appear to represent the decisive factor when a group of geographically dispersed computers are used to solve a large-scale problem.

Kalashnikova in 1979 used a similar problem to compare the basic iterative aggregation algorithms numerically with the variations and modifications of these algorithms presented in Sections 3.5 and 3.6 and with the algorithms for general linear systems described previously in this chapter. The variants that displayed the fastest rate of convergence were the block methods based on the Nekrasov algorithm, that is, (3.6.1)-(3.6.3), (3.6.11); and (3.5.2), (3.5.3), (3.6.11). These methods are 1.5 to 2.0 times faster than the corresponding nonblock methods. Algorithms (3.6.1)-(3.6.3), (3.6.8) and (3.5.2), (3.5.3), (3.6.9) appeared to be a little slower.

The rate of convergence of the basic one-level process (3.3.5), (3.3.6) has proven to be close to the rate of the block two-level process mentioned above. Of the specialized versions of the general one-level algorithm (3.3.5), (3.6.13), (3.4.14) with separate inter-industry and industry problems, the best rate of convergence has been found for those algorithms that use either the block Nekrasov algorithm or block simple iterations.

The experiments with the one- and two-level iterative aggregation algorithms have shown that if at each iteration the aggregate problem, rather than being solved exactly, is only approximated by one or two steps of successive approximations, the number of iterations needed to find an adequate solution does not increase by a significant amount. The numerical experiments with the many variants of the two-level process with minimization of imbalances (described in Section 3.5) have shown that the algorithms for solving systems of equations that perform the second stage using block methods have the fastest rate of convergence [the process described by (3.5.2), (3.5.3), (3.6.9) converges 1.5 times faster than does the process (3.5.2)-(3.5.4) that uses successive approximations]. Moreover, the rate of convergence of the processes with minimization

TABLE 4.1 Comparative Rate of Convergence of Iterative Aggregation Algorithms[1]

Number of algorithm	Name of algorithm	Numbers of formulas	Rate of convergence
1	Basic two-level algorithm	(3.6.1)-(3.6.4)	0,5
2	Two-level algorithm with Nekrasov method on the second stage of each iteration	(3.6.1)-(3.6.3), (3.6.8)	0,77
3	Two-level algorithm with block modification of Nekrasov method on the second stage of each iteration	(3.6.1)-(3.6.3), (3.6.11)	1
4	Two-level algorithm with minimization of imbalances in a semiaggregate problem and with successive approximation on the second stage of each iteration	(3.5.2)-(3.5.4)	0,5
5	Two-level algorithm with minimization of imbalances in a semiaggregate problem and with block modification of Nekrasov method on the second stage of each iteration	(3.5.2), (3.5.3), (3.6.11)	1
6	Two-level algorithm with minimization of imbalances in a semiaggregate problem and with block simple successive approximation on the second stage of each iteration	(3.5.2), (3.5.3), (3.6.9)	0,77

7	Two-level algorithm with minimization of imbalances both in a semiaggregate and in each intraindustry problem (intraindustry problems solved in parallel)	(3.5.2), (3.5.3), (4.1.4), (4.1.5)	0,06
8	The same with intraindustry problems solved sequentially	(3.5.2), (3.5.3), (4.1.6)	0,5
9	Group coordinate-wise descent		0,04
10	Two-level algorithm with ℓ_1-minimization of imbalances in a semiaggregate problem and with successive approximations on the second stage of each iteration	(3.5.16)-(3.5.21)	0,33
11	Two-level algorithm with minimization of imbalances in a semiaggregate problem and with aggregation of increments	(4.3.2)-(4.3.11)	0,17
12	One-level algorithm with minimization of imbalances in a semiaggregate problem and with aggregation of increments	(4.4.1), (4.4.2)	0,5

[1]Rates of convergence of various algorithms are compared with the algorithm no. 5, whose rate of convergence is taken equal to 1.

of imbalances which use block methods are similar to the rates of the two-level iterative aggregation algorithms (3.6.1)-(3.6.3), (3.6.11), which also use block methods.

Numerical studies using the generalized algorithms of this chapter have shown that they are not as effective at solving input-output models as are the algorithms described in Section 3.5. Thus the algorithm (3.5.2), (3.5.3), (4.1.4), (4.1.5) based on group coordinate descent methods, and hence averaging procedures, converges considerably more slowly than do the other two-level iterative aggregation algorithms that solve the minimization problem for the semiaggregate problem. For example, the rate of convergence is 16 times slower than that of the algorithm (3.5.2), (3.5.3), (3.6.11) using block Nekrasov methods. However, if we modify the general process by using the sequential algorithm (3.5.2), (3.5.3), (4.1.6) that uses group coordinate descent method at the industry level, the rate of convergence is increased by a factor of 6.

At the same time, when compared with their classical counterparts, the algorithms of this chapter perform favorably. For example, on the average, (3.5.2), (3.5.3), (4.1.4), (4.1.5) converges 1.5 times faster than does the usual method of group coordinate descent. The sequential algorithm (3.5.2), (3.5.3), (4.1.6) improves the convergence rate of the normal algorithm by a factor of 10.

The increment aggregation algorithm presented in Section 4.3 that computes the new approximation of problem (3.1.1) by

$$\begin{aligned} \hat{x}_g^\sigma &= x_g^{\sigma-1} + z_j^\sigma \Delta x_g^{\sigma-1}, \qquad g \in G_j,\ j = 1, \ldots, J; \\ x^\sigma &= A\hat{x}^\sigma + b, \qquad \hat{x}^\sigma = \{\hat{x}_g^\sigma \mid g = 1, \ldots, G\} \end{aligned} \tag{4.5.1}$$

converges at a rate of the same order as the regular algorithm (3.5.2)-(3.5.4), which uses minimization in the semiaggregate problem and successive approximations.

A summary of the evaluation of the two-level algorithms is given in Table 4.1. The table shows the rates of convergence of various algorithms compared with the best algorithms in the group, that is, the two-level algorithms that use block Nekrasov methods at the second stage. The table suggests the following conclusions: The table suggests the following conclusions:

1. As a rule, incorporating aggregation into any of the classical solution algorithms increases the rate of convergence by a factor of 1.5 to 2.0.
2. Aggregate problems as master problems do not affect the rate of convergence to a great extent.
3. For iterative aggregation algorithms in which aggregative transformations are incorporated into some successive approximation algorithm (e.g., successive approximations, simple iterations, Gauss-Seidel, Nekrasov iterations, and their block variants), it is sufficient when solving the aggregate problem to perform only one or two steps of the appropriate successive approximation algorithm.
4. Iterative aggregation algorithms based on successive approximation algorithms appear to be superior to those based on coordinate descent methods.
5. When aggregation transformations are built into successive approximation algorithms, the more effective the base algorithm is, the more effective is the iterative aggregation algorithm.

4.6 ITERATIVE AGGREGATION ALGORITHMS FOR OPTIMAL INTERPRODUCT INPUT-OUTPUT MODELS

By an "optimal input-output model" we mean a model in which the final consumption of goods and the volume of output need to be determined rather than being givens of the system. In addition, this model may include constraints on some or all of the primary scarce resources (e.g., production capacities, level of capital investment, size of labor force, etc.), as well as demand functions for consumer goods that are functions of personal income which is being maximized in the objective function of the problem. The model does retain the standard assumptions that there exists a single technology for the production of each commodity and the absence of so-called "multiple-output technologies" that would allow the same technology to produce two or more types of commodities. As a result, the matrix $(\hat{a}_{gq})$ of input-output relations is square.

The optimal input-output model is one of the simplest optimization models in economics and is the closest in form to a system of input-output equations. Formally, the model is described as

$$\Sigma_q \hat{a}_{gq} x_q + \hat{y}_g = x_g, \qquad g = 1, \ldots, G; \tag{4.6.1}$$

$$\hat{a}_{gq} \geqq 0; \tag{4.6.2}$$

$$\Sigma_g \hat{a}_{gq} < 1; \tag{4.6.3}$$

$$\hat{y}_g = \hat{y}_g(v), \qquad \frac{\partial}{\partial v} \sum_g \hat{y}_g(v) > 0; \tag{4.6.4}$$

$$\Sigma_q b_{sq} x_q \leqq R_s, \qquad b_{sq} \geq 0, \qquad s \in S; \tag{4.6.5}$$

$$\max v. \tag{4.6.6}$$

We shall also assume that $\lim_{v\to\infty} \Sigma_g \hat{y}_g(v) = \infty$. Here the subscript s denotes the primary scarce resources $(s \in S)$; v the consumer's disposable income, to be determined from the solution; $\hat{y}_g$ the volume of personal consumption of product g; $\hat{y}_g(v)$ the given function that relates the volume of personal consumption of product g to the consumer's disposable income; b_{sq} the given input coefficient of the sth primary scarce resource per unit of input of product q; and R_s the amount of the sth primary scarce resource available.

If we aggregate the variables x_g and the corresponding constraints (4.6.1) but not the constraints (4.6.5), the one- and two-level iterative aggregation algorithms for this case turn out to be similar to the algorithms discussed in Chapter 3 for the input-output equations. First the aggregate consumer demand functions for industry j's products are calculated as

$$y_i(v) = \sum_{g\in G_i} \hat{y}_g(v). \tag{4.6.7}$$

Assume that the approximate outputs $x_g^{\sigma-1}$ are given for iteration $\sigma - 1$. Then the next iteration proceeds as follows:

1. From (3.2.9) compute the input-output flows $X_{ij}^{\sigma-1}$ and the industries' output volumes $X_i^{\sigma-1}$, as well as the total amount of the scarce resources consumed by each industry:

$$B_{sj}^{\sigma-1} = \sum_{q \in G_j} b_{sq} x_q^{\sigma-1}. \qquad (4.6.8)$$

2. Solve the aggregate optimal input-output model:

$$\max v; \qquad (4.6.9)$$

$$X_i^{\sigma-1} z_i = \Sigma_j X_{ij}^{\sigma-1} z_i + y_i(v), \qquad i = 1, \ldots, J; \qquad (4.6.10)$$

$$\Sigma_j B_{sj}^{\sigma-1} z_j \leqq R_s, \qquad s \in S. \qquad (4.6.11)$$

The solution to this problem, denoted by z_j^{σ}, characterizes the changes in an industry's outputs in the new iteration as well as the new overall size v^{σ} of the consumer's disposable income.

3. Compute the new approximation:

$$x_g^{\sigma} = \sum_j \sum_{q \in G_j} \hat{a}_{gq}(z_j^{\sigma} x_q^{\sigma-1}) + \hat{y}_g(v^{\sigma}). \qquad (4.6.12)$$

When J = 1 the algorithm simplifies greatly. In Appendix 4.2 we prove that this algorithm coverges globally if J = 1 and the functions $\hat{y}_g(v)$ are close to linear functions. Before giving a detailed analysis of this algorithm we will show that the solution of (4.6.1)-(4.6.6) can be found if we know the solution of the following system for each ℓ ($1 \leq \ell \leq |S|$):

$$x_g = \Sigma_q \hat{a}_{gq} x_q + \hat{y}_g(v), \qquad g = 1, \ldots, G, \qquad (4.6.13)$$

$$R_\ell = \Sigma_q b_{\ell q} x_q. \qquad (4.6.14)$$

Let x(v) denote the solution of (4.6.13) for a given v. As we vary v the set of all x(v) defines a continuous curve in the space of vector x. Since we have assumed that the problem (4.6.1)-(4.6.6) has a solution, this curve must have at least one point of intersection with the polyhedron defined by (4.6.5). The conditions $b_{sq} \geqq 0$, $s \in S$, $q = 1, \ldots, G$, and $\lim_{v\to\infty} \Sigma_g \hat{y}_g(v) = \infty$ guarantee that this curve intersects all hyperplanes of constraints (4.6.5). Moreover, these conditions guarantee that the intersection of the curve and each of these hyperplanes is unique. This implies that one of the $|S|$ points of intersection of the curve $x(v)$ with the hyperplanes (4.6.5) is the solution of the problem (4.6.1)-(4.6.6), which is what we desired to show.

As $J = 1$, the problem (4.6.9)-(4.6.11) has only one z_j which is immediately derived from (4.6.11). Substituting this value into (4.6.10) we obtain the value of $\Sigma_g \hat{y}_g(v^\sigma)$. Since $\Sigma_g \hat{y}_g(v)$ is a monotonic function of v this allows us to solve for v^σ. Thus the algorithm in question is correctly defined.

Vakhutinsky, Dudkin, and Shchennikov (1973) proved the convergence of this algorithm under the assumption:

$$\left|\max_q(1 - \sum_g \hat{a}_{gq}) \frac{\partial(\hat{y}_g(v')/v)}{\partial v'} \frac{\partial v'}{\partial d_g}\right| < \hat{a}_{gq} + \frac{\hat{y}_g(v')}{v'} \times \min_q(1 - \sum_g \hat{a}_{gq}), \quad (4.6.15)$$

where $v' = \Sigma_g \hat{y}_g(v)$ and d_g is defined by (3.2.5). In Appendix 4.2 we show that given these assumptions $\hat{y}_g(v)$ can be regarded as a function of v'. Subsequently Rabinovich (1976) pointed out that this condition is satisfied by functions $\hat{y}_g(v')$, which are nearly linear, and if $J = 1$ and $S = 1$ the algorithm will converge globally. Indeed, if

$$\hat{y}_g(v') = c_g v' \quad (4.6.16)$$

then

$$\frac{\partial(\hat{y}_g(v')/v')}{\partial v'} = 0$$

and the condition (4.6.15) is satisfied. Although convergence proofs for this algorithm have been obtained only for very special assumptions, numerical experiments [Dyakonova and Novoselsky (1979)] have shown that the algorithm converges and even at a rapid rate for the general case, when $G > 1$.

Vakhutinsky, Grigorova, and Dudkin (1974) and Vakhutinsky, Dudkin et al. (1975) proposed a one-level iterative aggregation algorithm to solve (4.6.1)-(4.6.6). One iteration of the algorithm consists of the following steps.

(a) Industries compute the input-output flows, the industries' outputs, and the semiaggregate flows of primary resources corresponding to the approximate solution obtained at the previous iteration using (3.3.5), and pass the calculated volumes to each other.

(b) Each industry j solves a separate interproduct interindustry model:

$$x_g = \sum_{q \in G_j} \hat{a}_{gq} x_q + \sum_{r \neq j} \sum_{q \in G_r} \hat{a}_{gq} x_q^{\sigma-1} z_r^j + \hat{y}_g(v), \quad g \in G_j; \tag{4.6.17}$$

$$X_i^{\sigma-1} z_i^j = \sum_{q \in G_j} \tilde{\tilde{a}}_{iq} x_q + \sum_{r \neq j} X_{ir}^{\sigma-1} z_r^j + y_i(v), \qquad i \neq j; \tag{4.6.18}$$

$$\sum_{q \in G_i} b_{sq} x_q + \sum_{r \neq j} b_{sr}^{\sigma-1} z_r^j \leqq R_s, \qquad s \in S; \tag{4.6.19}$$

$$\max v, \tag{4.6.20}$$

where the subscript r denotes industry and $\tilde{\tilde{a}}_{iq} = \Sigma_{g \in G_i} \hat{a}_{gq}$.

Let x_g^σ, $g \in G_j$, $z_i^{j,\sigma}$, $i \neq j$ be the solution. Then x_g^σ, $g \in G_j$ for all j constitute the new approximation. These algorithms are the generalizations of (3.2.1)-(3.2.4) and (3.3.2)-(3.3.4) to the optimal input-output model. A generalization of the algorithm (3.5.2)-(3.5.4) can be constructed in a similar manner. However, for these latter algorithms convergence can not be achieved if the new approximation is found using (4.6.12). The reason for this is that for a matrix A with a sufficiently small norm, the transformation (4.6.12) decreases the imbalances in (4.6.13) but the norm of the total imbalances in (4.6.13) and (4.6.14) may increase. Therefore it is better to find the new approximation by minimizing the imbalances by changing the variables of each subsystem in turn.

Similarly, we can construct a two-level algorithm in which v^σ is calculated when the aggregate problem is solved; that is, we calculate simultaneously the values of v^σ and z_i^σ by minimizing

$$\sum_i \sum_{g \in G_i} \left(z_i x_g^{\sigma-1} - \sum_j \sum_{q \in G_j} \hat{a}_{gq} x_q^{\sigma-1} z_j - \hat{y}_g(v) \right)^2 + \left(\sum_j \sum_{q \in G_j} b_{sq} x^{\sigma-1} z_j - R_s \right)^2. \tag{4.6.21}$$

The ith subsystem then finds a solution for x_g by minimizing

$$\sum_{g\in G_i}\left\{x_g - \sum_{q\in G_i}\hat{a}_{gq}x_q - \sum_{j\neq i}\sum_{q\in G_j}\hat{a}_{gq}x_q^{\sigma-1}z_j^{\sigma} - \hat{y}_g(v^{\sigma})\right\}^2$$

$$+ \sum_{r\neq i}\sum_{g\in G_r}\left\{z_r^{\sigma}x_g^{\sigma-1} - \sum_{q\in G_i}\hat{a}_{gq}x_q - \sum_{j\neq i}\sum_{q\in G_j}\hat{a}_{gq}x_q^{\sigma-1}z_j^{\sigma} - \hat{y}_g(v^{\sigma})\right\}^2$$

$$+ \left(\sum_{i\neq j}\sum_{q\in G_j} b_{sq}x_q^{\sigma-1}z_j^{\sigma} + \sum_{q\in G_i} b_{sq}x_q - R_s\right)^2 \tag{4.6.22}$$

Finally, we can also calculate v^{σ} together with x_g^{σ}, that is, (4.6.22) is minimized with respect to x_q, $q \in G_i$, and v. In this case each industry has determined $\tilde{x}_q^{\sigma}$, $q \in G_i$, and a particular value of $v^{\sigma,i}$, and v^{σ} is computed as the average of these values

$$v^{\sigma} = \Sigma_i\, \alpha^{\sigma,i}v^{\sigma,i}, \qquad \alpha^{\sigma,i} \geqq \delta > 0, \qquad \Sigma_i\, \alpha^{\sigma,i} = 1;$$

and x_g^{σ} will be computed as:

$$x_g^{\sigma} = (1 - \alpha^{\sigma,i})z_i^{\sigma}x_g^{\sigma-1} + \alpha^{\sigma,i}\tilde{x}_g^{\sigma}, \qquad g \in G_i. \tag{4.6.23}$$

One-level algorithms can also be constructed which calculate v^{σ} in a similar fashion during the successive approximation step.

APPENDIX 4.1 ON THE CONVERGENCE OF ITERATIVE AGGREGATION ALGORITHMS FOR GENERAL LINEAR SYSTEMS OF EQUATIONS

In this appendix we prove the convergence properties for iterative aggregation algorithms for solving arbitrary systems of the form

$$Ax = b, \tag{A4.1.1}$$

described in this chapter. Detailed proofs will be presented only for the case where (A4.1.1) has a unique solution; the proof for the other cases will only be outlined. Thus unless otherwise stated, we assume that (A4.1.1) has a unique solution.

Before we prove convergence, we need to prove the following result:

LEMMA A4.1.1 Let $P(x)$ be a continuous operator such that $P(x^*) = x^*$, where x^* solves (A4.1.1) and the inequality $\|AP(x) - b\| < \|Ax - b\|$ holds for any $x \neq x^*$. Then the process $x^\sigma = P(x^{\sigma-1})$ converges to x^*.

Proof: Since we have assumed that (A4.1.1) has a unique solution the set of vectors x such that $\|Ax - b\| \leq c$ for any c forms an ellipsoid and is either bounded or empty. By assumption we have that if $x^\sigma = P(x^{\sigma-1})$, then $\|Ax^\sigma - b\| < \|Ax^{\sigma-1} - b\|$. It follows that for any x^0 the sequence $x^1, x^2, \ldots,$ lies in the ellipsoid $\|Ax - b\| \leq \|Ax^0 - b\|$. Hence the sequence has a limit point and we denote one such point by y. Since P is continuous by assumption, it follows that $\|AP(y) - b\| = \|Ay - b\|$. However, this implies that $y = x^*$. Hence any limit point of the sequence must equal x^*.

THEOREM A4.1.1 Two algorithms, determined by formulas (4.1.3) and (4.1.6), respectively, and the algorithm (4.2.1)-(4.2.3) converge to the solution x^* of (A4.1.1).

Proof: Direct computations show that the mappings $P\colon\ x^{\sigma-1} \to x^\sigma$ for each of the four algorithms are continuous and that $P(x^*) = x^*$. Hence from Lemma A4.1.1 the proof is complete if we can show that $\|Ax^\sigma - b\| < \|Ax^{\sigma-1} - b\|$.

Let us show that this condition is valid for the algorithm (4.1.3). Let $x^0 \neq x^*$ be an arbitrary vector $x^0 = \{\bar{x}_1^0, \ldots, \bar{x}_J^0\}$. Let $\hat{x}_i^1$ be the vector generated after we have solved the jth optimization problem. Then the next approximation would be determined from

$$x^1 = \gamma_1^0(\hat{x}_1^1, \bar{x}_2^0, \ldots, \bar{x}_J^0) + \gamma_2^0(\bar{x}_1^0, \hat{x}_2^1, \ldots, \bar{x}_J^0) + \cdots$$
$$+ \gamma_J^0(\bar{x}_1^0, \bar{x}_2^0, \ldots, \hat{x}_J^1), \qquad \gamma_j^0 > 0, \qquad \Sigma_j\ \gamma_j^0 = 1.$$

It is clear that for each term

$$\|\Lambda(\bar{x}_1^0, \ldots, \bar{x}_{i-1}^0, \hat{x}_i^1, \bar{x}_{i+1}^0, \ldots, \bar{x}_J^0)\| \leq \|\Lambda(x^0)\|,$$

where $\Lambda(x) = Ax - b$. We need to show that this inequality is strict for at least one of these terms. Consider the ellipsoid $\|\Lambda(x)\| \leq \|\Lambda(x^0)\|$. Let V_i denote the subspace $\{x_j = x_j^0,\ j \neq i\}$, which is nontangential to the given ellipsoid at the point x^0 (obviously not all of them can be tangential). We can select a vector $\hat{x}_i^1$ such that $(\bar{x}_1^0, \ldots, \hat{x}_i^1, \ldots, \bar{x}_J^0)$ lies inside the ellipsoid. From this it follows that

$$\|\Lambda(\bar{x}_1^0, \ldots, \hat{x}_i^1, \ldots, \bar{x}_J^0)\| < \|\Lambda(x^0)\|.$$

It is readily apparent that

$$\|\Lambda(x^1)\| \leqq \gamma_1^0 \|\Lambda(\hat{x}_1^1, \bar{x}_2^0, \ldots, \bar{x}_J^0)\| + \cdots$$
$$+ \gamma_J^0 \|\Lambda(\bar{x}_1^0, \ldots, \hat{x}_J^1)\| < \|\Lambda(x^0)\|.$$

This establishes the conditions of the lemma and the proof is complete.

Next we show that the assumption of the lemma is valid for the algorithm (4.2.3). Let $x^0 \neq x^*$ be an arbitrary vector $x^0 = (\bar{x}_1^0, \ldots, \bar{x}_J^0)$. Consider the subspace V_1 defined by $(x_j = x_j^0,\ j = 2, \ldots, J)$ and the lines $t\bar{x}_i^0$ $(i = 2, \ldots, J)$. If either the subspace or any of the lines is nontangential to the ellipsoid $\|\Lambda(x)\| \leqq \|\Lambda(x^0)\|$, then as in the proof of the convergence of the previous process, this implies that $\|\Lambda(\tilde{x}^1)\| < \|\Lambda(x^0)\|$, where $\tilde{x}^1 = (\tilde{x}_1^1, \tilde{x}_2^0, \ldots, \tilde{x}_J^0)$ denotes the point given by the solution of (4.2.1). Conversely, if the subspace and all the lines are tangential to the ellipsoid, the solution of the minimization problem would generate the original vector x^0. The same will be true for the subspace V_j and the lines $t\bar{x}_i^0$ $(i \neq j)$ since all the subspaces V_i, $i = 1, \ldots, J$, cannot be tangential to the ellipsoid. This implies that strict inequality must hold in at least one coordinate, which is the desired result.

A similar proof can be used to show that the algorithm (4.1.6) also satisfies the conditions of the lemma, and hence that they converge also. The following remarks can be made about these algorithms.

1. The proof is valid for any arbitrarily chosen vector norm and the corresponding matrix norm, where this is defined as

$$\|A\| = \sup_{\|x\|=1} \|Ax\|.$$

For this pair of norms it is true that

$$\|Ax\| \leq \|A\| \cdot \|x\|.$$

Theorem A4.1.1 requires that the norm be differentiable at all points except perhaps at 0.

2. Coordinate descent-based aggregation algorithms can be constructed for an arbitrary system Ax = b and convergence is guaranteed as long as the matrix A is nondegenerate.
3. Under the conditions of the theorem, if the system is degenerate, one can easily see that all the algorithms of this section satisfy $\lim_{\sigma \to \infty} \rho(x^\sigma, X^*) = 0$, where X* is the set of solution vectors of the problem of minimization of imbalances in the norm being used and

$$\rho(x^\sigma, X^*) = \inf_{x^* \in X^*} \|x^\sigma - x^*\|.$$

Now let us turn our attention to iterative incremental aggregation algorithms for solving (A4.1.1).

THEOREM A4.1.2 If (A4.1.1) has a unique solution, the algorithm (4.3.2)-(4.3.10) as well as the sequential algorithm (4.4.1), (4.4.2) and its parallel counterpart all converge using either method to determine the variable increments described in Section 4.3.

The proof of Theorem A4.1.2 follows directly from Lemma A4.1.1. All the comments made in remark 3 apply equally to these algorithms.

APPENDIX 4.2 MATHEMATICAL ANALYSIS OF THOSE ITERATIVE AGGREGATION ALGORITHMS FOR THE OPTIMAL INTERPRODUCT INPUT-OUTPUT MODEL WHICH USE THE INTERINDUSTRY PROBLEM

We will prove the convergence of the algorithm (4.6.8)-(4.6.12) when J = 1 and when the functions $\hat{y}_g(v)$ are all close to linear. As noted previously, the problem (4.6.1)-(4.6.6) is equivalent to the problem (4.6.13)-(4.6.14). Thus the algorithm (4.6.8)-(4.6.12) reduces to first finding z^σ from $x^{\sigma-1}$:

$$z^\sigma = \frac{R_\ell}{\Sigma_q b_{\ell q} x_q^{\sigma-1}}; \qquad \text{(A4.2.1)}$$

then calculating v^σ from (4.6.13):

$$\Sigma_g \hat{y}_g(v) = z^\sigma \Sigma_g x_g^{\sigma-1} - z^\sigma \Sigma_g \Sigma_q \hat{a}_{gq} x_q^{\sigma-1} \quad \text{(A4.2.2)}$$

and computing the new approximation by

$$x_g^\sigma = z^\sigma \Sigma_q \hat{a}_{gq} x_q^{\sigma-1} + \hat{y}_g(v^\sigma). \quad \text{(A4.2.3)}$$

If we strenghthen the assumptions slightly so that

$$\sum_g \frac{\partial \hat{y}_g(v)}{\partial v} > 0$$

for all v, then (A4.2.2) has a unique solution.

We now show that we can assume in the process described above that

$$\Sigma_g \hat{y}_g(v) = v$$

by some change of variable v, namely,

$$v' = \Sigma_g \hat{y}_g(v),$$

we clearly have

$$\frac{\partial v'}{\partial v} > 0, \qquad \frac{\partial \hat{y}_g(v')}{\partial v'} > 0.$$

Therefore the original problem can be rewritten in terms of v' with the same constraints.

We shall assume that $v = \Sigma_g \hat{y}_g(v)$ for all v. Let d_g^σ denote $x_g^\sigma/\Sigma_q x_q^\sigma$. Then as in the proof of the algorithm (3.2.1)-(3.2.4) when J = 1, we can show (Vakhutinsky, Dudkin, and Shchennikov, 1973) that

$$d_g^\sigma = \Sigma_q \left[\hat{a}_{gq} + \frac{\hat{y}_g(v^\sigma)}{v^\sigma} (1 - \Sigma_t \hat{a}_{tq})\right] d_q^{\sigma-1}. \quad \text{(A4.2.4)}$$

We will prove that if

$$\left|\max_k(1 - \Sigma_t \hat{a}_{tk}) \frac{\partial \tau_g(v)}{\partial v} \frac{\partial v}{\partial d_q}\right| < \hat{a}_{gq} + \tau_g(v) \min_k(1 - \Sigma_t \hat{a}_{tk}), \quad \text{(A4.2.5)}$$

where

$$\tau_g(v) = \frac{\hat{y}_g(v)}{v}$$

is valid for all g and q, then the algorithm (4.6.8)-(4.6.11) converges. (The partial derivatives $\partial v/\partial d_g$ have the following meaning. The new value of v determined from the aggregate model (4.6.9)-(4.6.11) is, in fact, a function of aggregation weights satisfying condition $\Sigma_g d_g = 1$. This function is extended to the entire space without the restriction $\Sigma_g d_g = 1$ by the requirement that it be homogeneous of degree 1.)

Conditions (A4.2.5) are satisfied, for example, if $y_g(v)$ are linear functions. Indeed, if

$$\hat{y}_g(v) = c_g v,$$

where c_g is a constant, we have

$$\frac{\partial \tau_g(v)}{\partial v} \frac{\partial v}{\partial d_q} = \frac{\partial \tau_g(v)}{\partial d_q} = 0.$$

Since

$$\tau_g(v) \min_q \Sigma_q \hat{a}_{gq} \geq 0, \qquad \hat{a}_{gq} \geq 0,$$

then conditions (A4.2.5) are satisfied.

Now we can formulate

THEOREM A4.2.1 If the conditions (A4.2.5) are valid then the algorithm (4.6.8)-(4.6.11) with J = 1 converges to the solution for any starting approximation.

Proof: It is straightforward to show that the solution of (4.6.13), (4.6.14) is a stationary point of the algorithm. As a contraction mapping has a unique fixed point, in order to prove the theorem it will be sufficient to show that the mapping defined by (A4.2.4) is contracting, as it clearly maps the unit simplex into itself. We will make use of the fact that $x^{\sigma+1}$ is uniquely determined from the vector d^{σ}.

The Jacobian of the mapping (A4.2.4) at an arbitrary point d^σ is:

$$\frac{\partial d_g^{\sigma+1}}{\partial d_q^\sigma} = \hat{a}_{gq} + \tau_g(v^{\sigma+1})(1 - \Sigma_t \hat{a}_{tq})$$

$$+ \frac{\partial \tau_g(v^{\sigma+1})}{\partial d_q^\sigma} \Sigma_k d_k^\sigma(1 - \Sigma_t \hat{a}_{tk}). \qquad \text{(A4.2.6)}$$

The proof will proceed by showing that at any point the Jacobian of the mapping is a positive stochastic matrix. It is a positive matrix since:

$$\left|\frac{\partial \tau_g(v^{\sigma+1})}{\partial d_q^\sigma} \sum_k d_k^\sigma(1 - \Sigma_t \hat{a}_{tk})\right| \le \left|\max_k(1 - \Sigma_t \hat{a}_{tk}) \frac{\partial \tau_g(v^{\sigma+1})}{\partial d_q^\sigma}\right|$$

$$< \hat{a}_{gq} + \tau_g(v^{\sigma+1}) \min_k(1 - \Sigma_t \hat{a}_{tk})$$

$$\le \hat{a}_{gq} + \tau_g(v^{\sigma+1})(1 - \Sigma_t \hat{a}_{tq})$$

[this is true by (A4.2.5)] and the column sums are all equal to 1:

$$\Sigma_g\left(\hat{a}_{gq} + \tau_g(v^{\sigma+1})(1 - \Sigma_k \hat{a}_{kq}) + \frac{\partial \tau_g(v^{\sigma+1})}{\partial d_q^\sigma} \sum_k d_k^\sigma(1 - \Sigma_t \hat{a}_{tk})\right)$$

$$= \Sigma_g \hat{a}_{gq} + (1 - \Sigma_k \hat{a}_{kq}) \Sigma_g \tau_g(v^{\sigma+1}) + \Sigma_k (1 - \Sigma_t \hat{a}_{tk}) d_k^\sigma \sum_g \frac{\partial \tau_g(v^{\sigma+1})}{\partial d_q^\sigma}$$

$$= \Sigma_g \hat{a}_{gq} + 1 - \Sigma_k \hat{a}_{kq} = 1,$$

as $\Sigma_g \tau_g(v) = 1$ for all v. Thus the Jacobian of the map (A4.2.4) is a positive, stochastic matrix at any point d^σ and the entries of the Jacobian depend continuously on d^σ. Lubitch (1974) has shown that the norm of the stochastic matrix $S = (s_{gq})$ on the unit simplex is

$$\frac{1}{2} \max_{q,k} \sum_{g} |s_{gq} - s_{gk}|. \tag{A4.2.7}$$

As the Jacobian is continuous in d^σ there exists a $\gamma > 0$ such that all the entries of the Jacobian are greater than γ. From (A4.2.7) we have that the map defined by (A4.2.4) is locally contracting at any point d^σ and that in the ℓ_1 norm the contraction coefficient is not larger than $1 - 1/2\,\gamma + \varepsilon$. However, a locally contracting map defined on a simply connected compact set in a Euclidean space is globally contracting, and hence has a unique fixed point.

APPENDIX 4.3 MATHEMATICAL ANALYSIS OF THOSE ITERATIVE ALGORITHMS FOR THE OPTIMAL INTERPRODUCT INPUT-MODEL WHICH USE MINIMIZING IMBALANCES IN A SEMIAGGREGATE PROBLEM

To analyze the algorithms for solving the problem:

$$\max v, \tag{A4.3.1}$$

$$Ax + y(v) = x, \tag{A4.3.2}$$

$$B_1 x = c_1, \tag{A4.3.3}$$

$$B_2 x \leqq c_2, \tag{A4.3.4}$$

we introduce the function

$$F(x,v) = (Ax + y(v) - x)^2 + (B_1 x - c_1)^2 + (B_2 x - c_2)^2_{(+)}. \tag{A4.3.5}$$

As noted previously, if

$$\partial y_g(v)/\partial v \geqq 0, \qquad g = 1, \ldots, G, \tag{A4.3.6}$$

$$B_1 \geqq 0, \qquad B_2 \geqq 0,$$

then (A4.3.1)-(A4.3.4) has the same solution as the problem that minimizes $F(x,v)$ with respect to x and v. As the algorithms (4.6.21)-(4.6.23) are continuous maps, the proof is obtained by showing that the vector (x^*,v^*) represents the only stationary point of these maps and that the sequence (x^0,v^0), (x^1,v^1), ... converges to this point.

The algorithms (4.6.21)-(4.6.23) are somewhat apart from these algorithms. As distinct from the theorem (A4.1.1), the function F(x,v) in this instance is still convex with respect to x but may be concave with respect to v. However, even in this case if the functions $\hat{y}_g(v)$, g = 1, ..., G, are twice differentiable in a neighborhood of (x*,v*) and for at least one g, $\hat{y}_g(\cdot)$ is strictly increasing, i.e.,

$$\frac{\partial \hat{y}_g(v^*)}{\partial v} > 0$$

then there will exist a neighborhood of (x,v*) in which F(x,v) is convex in both variables. The next lemma proves that in a neighborhood of v*, F(x,v) is locally convex with respect to v.

LEMMA A4.3.1 There exists a neighborhood of the optimal point (x*, v*) of (A4.3.1)-(A4.3.5) in which F(x,v) is convex with respect to v.

Proof: There exists some neighborhood of point (x*,v*) such that for all (x,y) ∈ U we have:

$$\frac{\partial^2 F(x,v)}{\partial v^2} \geq 0.$$

But

$$\frac{\partial^2 F}{\partial v^2} = \frac{\partial^2 (Ax + y(v) - x)^2}{\partial v^2} = \frac{\partial\left(-2((I - A)x - y(v))\, \frac{\partial y}{\partial v}\right)}{\partial v}$$

$$= -2\left[(I - A)x\, \frac{\partial^2 y}{\partial v^2} - \left(\frac{\partial y}{\partial v}\right)^2 - y(v)\, \frac{\partial^2 y}{\partial v^2}\right]$$

$$- 2\left[\frac{\partial^2 y}{\partial v^2}\, ((I - A)x - y(v))\right],$$

where I is a unit G × G matrix. Since $\partial\hat{y}_g/\partial v$ is continuous for all g and $(\partial y/\partial v)^2 = \Sigma_g(\partial\hat{y}_g/\partial v)^2$ it follows that there exists a neighborhood of v* such that $(\partial y/\partial v)^2 > 0$. On the other hand, since there exists a neighborhood of v* in which $\partial^2\hat{y}_g/\partial v^2 < M < \infty$, g = 1, ..., G, and the term (I - A)x - y(v) can be made arbitrarily small by choosing a sufficiently small neighborhood, it follows that we can find a neighborhood in which $\partial^2 F/\partial v^2 > 0$.

PART III

ITERATIVE AGGREGATION ALGORITHMS FOR SOLVING EXTREMUM PROBLEMS

5

Iterative Aggregation Methods for Solving Unconstrained Optimization Problems

5.1 TWO-LEVEL ITERATIVE AGGREGATION ALGORITHMS

In this chapter we develop a general theory of iterative aggregation for unconstrained optimization problems. These problems consist of finding an extremum (maximum) of a function of a finite number of variables over the whole of $\mathbb{R}^K$, that is,

$$\max f(x); \tag{5.1.1}$$

$$x \in \mathbb{R}^K, \tag{5.1.2}$$

where $x = (\xi_1,\ldots,\xi_K)$ is a vector in $\mathbb{R}^K$. Unconstrained optimization problems provide the best setting for developing iterative aggregation algorithms for problems with aggregation of variables.

As noted before, iterative aggregation theory deals with solving an original problem by solving a sequence of problems for some "upper-level" variables such that each of them characterizes a whole group of the original variables. Also, there must be a method for calculating the values of the original variables having obtained the values of the upper-level variables from the aggregate problem. This means that it is necessary to construct K functions of J variables $\varphi_k(z_1,\ldots,z_J)$ which map the aggregate values into the original variables, that is,

$$\xi_k = \varphi_k(z_1,\ldots,z_J), \qquad k = 1, \ldots, K. \tag{5.1.3}$$

Using (5.1.3), we can vary the values of the aggregate variables in order to find a J-dimensional vector $z = z^\sigma$ at iteration σ that optimizes an appropriate objective function, where $z = (z_1,\dots,z_J)$. A natural candidate for this objective function is $f(\varphi(z))$. This is equivalent to making a change of variables in (5.1.1) using (5.1.3) to obtain

$$\max f(\varphi(z)); \tag{5.1.4}$$

$$z \in \mathbb{R}^J, \tag{5.1.5}$$

which we shall refer to as the aggregate problem.

We will confine our analysis to linear transformations of variables; that is, (5.1.3) will be of the form

$$\xi_k = b_k + \sum_{j=1}^{J} \varphi_{kj} z_j, \qquad k = 1, \dots, K. \tag{5.1.6}$$

We will use the following terminology to differentiate various types of aggregation. By "general linear aggregation" we mean aggregation according to (5.1.6). If $b_k = 0$ for all k in (5.1.6), we call this linear proportional aggregation. Suppose that both the index set of the original variables (i.e., $\{1,\dots,K\}$) and the index set of the aggregate variables (i.e., $\{1,\dots,J\}$) are partitioned into N nonempty disjoint subsets K and J, respectively; that is,

$$\bigcup_1^N K_n = \{1,\dots,K\}, \qquad \bigcup_1^N J_n = \{1,\dots,J\},$$

$$K_n \cap K_{n'} = \emptyset, \qquad J_n \cap J_{n'} = \emptyset \qquad \text{for } n \neq n'.$$

Let the matrix (φ_{kj}) consist of blocks whose rows correspond to the partitioning of the original variables and whose columns correspond to the partition of the aggregated variables; therefore, $\varphi_{kj} = 0$ if $k \in K_n$ and $j \in J_{n'}$, and $n \neq n'$.

Aggregation using this matrix φ will be referred to as linear block aggregation. Simple aggregation is a special case of linear block aggregation when the number of blocks is equal to J; that is, each block of the matrix φ has exactly one column. Note that in simple aggregation we fix the proportions between variables in each block. In what follows we omit the word "linear" in our definition of general aggregation, as our analysis will be restricted to this type of aggregation.

The aggregate problem (5.1.4), (5.1.5) is not equivalent to the original problem (5.1.1), (5.1.2) but rather is only an approximate description of the original problem. This means that it is necessary to examine the sequence of problems

$$f(\varphi^{\sigma}(z)) \to \max; \tag{5.1.7}$$

$$z \in \mathbb{R}^{J}. \tag{5.1.8}$$

Let z^{σ} denote the solution to (5.1.7), (5.1.8). In constructing the sequence of problems it is necessary that the sequence $x^{\sigma} = \varphi^{\sigma}(z^{\sigma})$ converge to the solution of (5.1.1), (5.1.2).

A highly desirable property of any maximization algorithm is nondecreasing of the objective function on the sequence of approximations x^{σ}; that is, $f(x^{\sigma-1}) \leqq f(x^{\sigma})$. This condition will necessarily be true if $x^{\sigma-1}$ is a feasible point for the aggregate problem at iteration σ, that is, if there exists an aggregate vector $z = e^{\sigma}$ such that $x^{\sigma-1} = \varphi^{\sigma}(e^{\sigma})$. This is so because at iteration σ we solve the problem $f(\varphi^{\sigma}(z)) \to \max$ and z^{σ} is the solution and $f(\varphi^{\sigma}(z^{\sigma})) \geq f(\varphi^{\sigma}(e^{\sigma})) = f(x^{\sigma-1})$. Thus since $x^{\sigma} = \varphi^{\sigma}(z^{\sigma})$, we have $f(x^{\sigma}) \geq f(x^{\sigma-1})$.

In many applications it is desirable to have the vector e^{σ} (that produces $x^{\sigma-1}$ as the feasible point) identically equal to 0 and to shift the origin of the aggregate variables. In this instance (5.1.6) becomes

$$\xi_k = \varphi^{\sigma}(z) = \xi_k^{\sigma-1} + \sum_{j=1}^{J} \varphi_{kj}^{\sigma} z_j, \qquad k = 1, \ldots, K. \tag{5.1.9}$$

For the proportional aggregation it is convenient to set $e_j^{\sigma} = 1$ and to adjust (5.1.6) appropriately. For simple aggregation this yields

$$\xi_k = \xi_k^{\sigma-1} z_j \qquad \text{for } k \in K, \qquad j = 1, \ldots, J. \tag{5.1.10}$$

We will confine our analysis to those iterative aggregation algorithms that determine the aggregation parameters in (5.1.6) from the approximate solution $x^{\sigma-1}$ obtained at iteration $\sigma - 1$, that is, $b_k^\sigma = b_k(x^{\sigma-1})$, $\varphi_{kj}^\sigma = \varphi_{kj}(x^{\sigma-1})$. With this restriction the function $\varphi^\sigma(z)$ is uniquely determined by $x^{\sigma-1}$, that is, $\varphi^\sigma(z) = \varphi(x^{\sigma-1}, z)$, and hence the solution z^σ of (5.1.7), (5.1.8) will also be a function of $x^{\sigma-1}$, that is, $z^\sigma = z(x^{\sigma-1})$.

When the general iterative aggregation algorithm (5.1.7)-(5.1.9) is compared with the majority of well-known methods for solving unconstrained optimization problems (such as feasible directions, quasi-Newton, conjugate gradient, etc.), it can be seen that the latter methods are special cases of the more general iterative aggregation algorithm with the dimension J of the aggregate problem equal to 1.

The following geometric interpretation of aggregation will make the discussion to follow more intuitive. The vectors $(\varphi_1(z), \ldots, \varphi_K(z))$ define a subset T in $\mathbb{R}^K$. From this fact and from our preceding remarks we can see that the aggregate problem solves the original problem over a subset of the domain of the problem. That is, the aggregate problem is derived from the original problem by restriction of the original problem onto the subset. This may give the impression that the aggregate problem has needlessly complicated the original problem. However, since the subset T can be defined as $\varphi(\mathbb{R}^J)$, the variables $z_1, \ldots, z_J$ define a coordinate system on this subset. Also, since f(x) is a composite function of $z_1, \ldots, z_J$, the aggregate problem is the one of maximization of $f(\varphi^\sigma(z))$ over $z \in \mathbb{R}^J$.

From this perspective the various aggregation algorithms differ only in their choice of the subset T. When using simple aggregation T is a cone spanned by the vectors $x_j^{\sigma-1}$. In this case, if ξ_k^σ is assumed to equal $\varphi_k(z^\sigma)$, the algorithm will a priori not converge. This was the case for most of the original iterative aggregation algorithms. One method for overcoming this problem (which was originally devised for solving constrained optimization problems) is to use an aggregate problem to adjust a regular iterative algorithm of the form

$$x^\sigma = \Psi(x^{\sigma-1}). \tag{5.1.11}$$

Thus one iteration requires two steps. In the first step we obtain the vector $\tilde{x}^{\sigma}$ from the vector $x^{\sigma-1}$ by $\tilde{x}^{\sigma} = \varphi^{\sigma}(z^{\sigma}) = \varphi(x^{\sigma-1}, z(x^{\sigma-1}))$, and at the second step we apply the iterative process Ψ: $x^{\sigma} = \Psi(\tilde{x}^{\sigma})$.

This procedure must be followed when using simple aggregation (5.1.10), which fixes the proportions between variables in a subsystem (group). Then one possible algorithm will proceed as follows. Assume that we are given the approximation $x^{\sigma-1}$. Using the aggregating functions

$$\varphi_k^{\sigma}(z) = \xi_k^{\sigma-1} z^j, \qquad k \in K_j, \qquad j = 1, \ldots, J,$$

construct the aggregate objective function

$$f^{\sigma}(z) = f(\varphi^{\sigma}(z)) \tag{5.1.12}$$

and solve the aggregate problem

$$\max f^{\sigma}(z); \qquad z \in \mathbb{R}^J.$$

Having found the solution z^{σ} we adjust the approximation by

$$\tilde{x}_k^{\sigma} = x_k^{\sigma-1} z_j^{\sigma}, \qquad k \in K_j, \qquad j = 1, \ldots, J.$$

Finally, we derive x^{σ} from $\tilde{x}^{\sigma}$ by performing one step of the group coordinate descent method on the function $f(x)$. If the set of solutions to the original problem is bounded and $f(x)$ is a smooth concave function, this algorithm will converge to the optimal solution from any initial approximation. In fact, this algorithm will converge for any choice of Ψ that monotonically converges with respect to some functional.

Computationally more effective algorithms can be obtained by using general linear aggregation instead of simple aggregation. This allows for much greater freedom in the choice of the subset T. When using general linear aggregation, any convergent iterative solution algorithm that incorporates aggregation to adjust the solution will also converge. However, a careful choice of T can obviate the need for the adjustment. When general iterative aggregation algorithms are used to generalize classical methods to the case of $J \geq 1$, we are able to use the sophisticated results about these methods to prove similar results about the iterative aggregation algorithms.

In the remainder of this section we use this approach to examine block aggregation in which each aggregate variable z_j corresponds to one group of variables K_j (where the number of groups of variables is $J \geqq 1$). In this case $\varphi_{kj}^{\sigma} = 0$ if $k \notin K_j$; that is, aggregation is done by fixing the proportional increments between variables in each group. Consider the vector $\Delta x^{\sigma-1} = (\Delta\xi_k^{\sigma-1} | k = 1, \ldots, K)$, where $\Delta\xi_k^{\sigma-1}$ stands for φ_{kj}^{σ}, $k \in K_j$. If we use the gradient of $f(x)$ at $x^{\sigma-1}$ for $\Delta x^{\sigma-1}$,

$$\Delta x^{\sigma-1} = \nabla f(x^{\sigma-1}), \tag{5.1.13}$$

we obtain a generalization of the classical gradient method. If the Hessian matrix $\nabla_2 f(x)$ is negative definite at $x^{\sigma-1}$, we choose $\Delta x^{\sigma-1}$ to be

$$\Delta x^{\sigma-1} = -[\nabla_2 f(x^{\sigma-1})]^{-1} \nabla f(x^{\sigma-1}). \tag{5.1.14}$$

Equation (5.1.14) generalizes Newton's method to the case where $J > 1$. Newton's method can also be generalized by using the quadratic approximation to the "truncated" function $f_j^{\sigma}(x_j) = f(x_1^{\sigma-1}, \ldots, x_{j-1}^{\sigma-1}, x_j, x_{j+1}^{\sigma-1}, \ldots, x_J^{-1})$, where $x_j = (\xi_k | k \in K_j)$ is the jth subvector of x, for computing the jth subvector $\Delta x_j^{\sigma-1} = (\Delta\xi_k^{\sigma-1} | k \in K_j)$ of $\Delta x^{\sigma-1}$. In this case

$$\Delta x_j^{\sigma-1} = -[\nabla_2 f_j^{\sigma}(x^{\sigma-1})]^{-1} \nabla f_j^{\sigma}(x^{\sigma-1}), \qquad j = 1, \ldots, J. \tag{5.1.15}$$

The matrix $\nabla_2 f_j^{\sigma}(x^{\sigma-1})$ is the $(K_j \times K_j)$-dimensional submatrix of the Hessian matrix of the function $f(x)$ evaluated at the point $x^{\sigma-1}$. Equation (5.1.15) requires that each of these submatrices be negative definite for $j = 1, \ldots, J$; this, however, is a weaker condition than requiring that the entire Hessian matrix be negative definite.

When one of the last three equations is used to calculate $\Delta x^{\sigma-1}$ and the function $f(x)$ is convex and has a bounded set of optima, the algorithm will converge for any initial value. In addition, in these algorithms it is not necessary to use an additional transformation Ψ (5.1.11). Next we give a more detailed analysis of the algorithm.

1. The VDI $\Delta x^{\sigma-1}$ is computed from any of the equations (5.1.13)-(5.1.15) evaluated at the point $x^{\sigma-1}$.
2. If the norm of $\Delta x^{\sigma-1}$ is sufficiently small, $x^{\sigma-1}$ is used as the solution.
3. Otherwise, construct the aggregate function

$$f^{\sigma}(z) = f(x^{\sigma-1} + \Sigma_j z_j \Delta x_j^{\sigma-1}). \tag{5.1.16}$$

4. Solve the aggregate problem

$$f^{\sigma}(z) \to \max; \tag{5.1.17}$$

$$z \in \mathbf{R}^J \tag{5.1.18}$$

and denote the solution by z_j^{σ}.

5. Compute the vector $x^{\sigma} = (\xi_1^{\sigma}, \ldots, \xi_K^{\sigma})$ from

$$\xi_k^{\sigma} = \xi_k^{\sigma-1} + \Delta\xi_k^{\sigma-1} z_j^{\sigma}, \qquad k \in K_j, \qquad j = 1, \ldots, J. \tag{5.1.19}$$

These iterative aggregation algorithms for unconstrained optimization can be seen to be generalizations of relaxation methods with one-dimensional optimization. Clearly, it makes sense to apply these algorithms only if the aggregated low-dimensional problem is not too much more complicated than one-dimensional, and this is often the case in the economic models. At the same time multidimensional optimization on each iteration seems to be preferable than one-dimensional because in the real cases the information is usually spread over different subsystems and exchange of information may take more time than the computations themselves. Therefore, it would seem preferable to reduce the total number of iterations while increasing the amount of computation performed for each iteration.

5.2 ONE-LEVEL ITERATIVE AGGREGATION ALGORITHMS FOR UNCONSTRAINED OPTIMIZATION PROBLEMS

The general parallel one-level iterative aggregation algorithm proceeds as follows. Suppose that at iteration σ each subsystem j has the approximation vector $x^{\sigma-1}$ and R_j vectors that have as their coordinates the increments $\Delta x_1^{j,\sigma-1}, \ldots, \Delta x_{R_j}^{j,\sigma-1}$. Using these vectors subsystem j solves the following local problem:

$$\max \hat{f}_j^{\sigma}(z) = f\left(x^{\sigma-1} + \sum_{i=1}^{R_j} z_i^j \Delta x_i^{j,\sigma-1}\right); \tag{5.2.1}$$

$$z^j \in \mathbb{R}^{R_j}, \tag{5.2.2}$$

whose optimal solution is denoted by $z^{j,\sigma}$. After this we compute the preliminary approximation $\tilde{x}^{j,\sigma}$:

$$\tilde{x}^{j,\sigma} = x^{j,\sigma-1} + \sum_{i=1}^{R_j} z_i^{j,\sigma} \Delta x_i^{j,\sigma-1}. \tag{5.2.3}$$

The final approximation is then found as

$$x^{\sigma} = \sum_{j=1}^{J} \gamma_j^{\sigma} \tilde{x}^{j,\sigma}; \qquad \sum_j \gamma_j^{\sigma} = 1,$$
$$\gamma_j^{\sigma} \geqq \delta > 0, \qquad j = 1, \ldots, J. \tag{5.2.4}$$

In the sequential one-level algorithm the subsystems solve the problem (5.2.1), (5.2.2) sequentially. Subsystem j constructs the vectors $x^{j,\sigma-1}, \Delta x_1^{j,\sigma-1}, \ldots, \Delta x_{R_j}^{j,\sigma-1}$ using the results from iteration σ for subsystems 1 through j - 1. In the simplest one-level algorithms, the VDIs $\Delta x_i^{j,\sigma}$ are constructed as follows. For subsystem j there are $J - 1 + |K_j|$ VDIs. The first J - 1 of these, denoted by $\Delta x_i^{j,\sigma-1}$, $i \neq j$, correspond to increments in the variables of the other subsystems. They are obtained from a single increment vector $\Delta x^{\sigma-1}$ by taking its K_i subvectors† $\Delta x_i^{\sigma-1}$:

---

†K_i subvectors v_i of vector $v \in \mathbb{R}^K$, $v = (v_k | k \in K)$, are defined as the projection of v onto $\mathbb{R}^{K_i}$: $v_i = (v_k | k \in K_i)$.

$$\Delta x^{\sigma-1} = (\Delta x_i^{\sigma-1} | i = 1, \ldots, J),$$

$$\Delta x_i^{j,\sigma-1} = \Delta x_i^{\sigma-1}, \qquad i \neq j.$$

(Thus $\Delta x_i^{j,\sigma-1}$ is independent of j if $j \neq i$.)

The other $|K_j|$ vectors are the unit coordinate vectors e_k, $k \in K_j$ and $e_k = 0$, $k \notin K_j$. This method of constructing the VDIs means that subsystem j does not aggregate its own variables in constructing its aggregate problem. Using standard notation let x_j^σ denote the K_jth subvector of the vector x^σ; that is, $x_j^\sigma = (\varepsilon_k^\sigma | k \in K_j)$.

Iteration σ of the simplest parallel one-level algorithm proceeds as follows.

1. Determine $\Delta x^{\sigma-1}$ from any of the formulas (5.1.13)-(5.1.15). If the norm of the vector is sufficiently small, $x^{\sigma-1}$ is the final solution.
2. For i = 1, ..., J, construct the vector $\Delta x_i^{\sigma-1}$ as the K_i subvector of the vector $\Delta x^{\sigma-1} = (\Delta\xi_1^{\sigma-1}, \ldots, \Delta\xi_K^{\sigma-1})$, that is, $\Delta x_i^{\sigma-1} = (\Delta\xi_k^{\sigma-1} | k \in K_i)$.
3. In each subsystem j solve the aggregate problem with unknowns z_i^j $(i \neq j)$ and the $|K_j|$-dimensional vector x_j:

$$f(x_1^{\sigma-1} + z_1^j \Delta x_1^{\sigma-1}, \ldots, x_{j-1}^{\sigma-1} + z_{j-1}^j \Delta x_{j-1}^{\sigma-1},$$
$$x_j, x_{j+1}^{\sigma-1} + z_{j+1}^j \Delta x_{j+1}^{\sigma-1}, \ldots, x_J^{\sigma-1} + z_J^j \Delta x_J^{\sigma-1}) \to \max. \tag{5.2.5}$$

Let $z_i^{j,\sigma}$, $\tilde{x}_j^\sigma$ denote the solution to (5.2.5).

4. Find the preliminary approximation:

$$\tilde{x}^{j,\sigma} = (x_1^{\sigma-1} + z_1^{j,\sigma} \Delta x_1^{\sigma-1}, \ldots, x_{j-1}^{\sigma-1} + z_{j-1}^{j,\sigma} \Delta x_{j-1}^{\sigma},$$
$$\tilde{x}_j^\sigma, x_{j+1}^{\sigma-1} + z_{j+1}^{j,\sigma} \Delta x_{j+1}^{\sigma-1}, \ldots, x_J^{\sigma-1} + z_J^{j,\sigma} \Delta x_J^{\sigma-1}). \tag{5.2.6}$$

5. Compute $x^{j,\sigma}$ from (5.2.4).

In the sequential algorithm the subsystem j uses the results from the most recent computations in order to form its aggregate problem. The order of the subsystems is arbitrary, but if it coincides with the way they are numbered, subsystem j's problem for the simplest algorithm becomes

$$f(x_1^{j,\sigma} + z_1^j \Delta x_1^{j,\sigma}, \ldots, x_{j-1}^{\sigma} + z_{j-1}^j \Delta x_{j-1}^{j,\sigma},$$
$$x_j, x_{j+1}^{j,\sigma-1} + z_{j+1}^j \Delta x_{j+1}^{j,\sigma-1}, \ldots, x_J^{\sigma-1} + z_J^j \Delta x_J^{j,\sigma-1}) \to \max. \tag{5.2.7}$$

Then preliminary approximation $\tilde{x}^{j,\sigma}$ is found according to (5.2.6) and then it is conveyed to the next subsystem:

$$x^{j+1,\sigma-1} = \tilde{x}^{j,\sigma} \quad \text{if } j < J;$$
$$x^{1,\sigma} = \tilde{x}^{J,\sigma}.$$

For this algorithm it is not necessary to average the x^{σ} using (5.2.4) since x^{σ} is already composed of the subvectors x_j^{σ}. The relevant VDIs $\Delta x^{j,\sigma}$ are determined as in the parallel algorithms.

The parallel one-level algorithms converge at a slower rate than do the sequential algorithms. However, the parallel algorithms solve the subsystems' problems concurrently, which is more reflective of real-time problems in which the necessary information is concentrated in different subsystems.

The simplest one-level algorithm is convergent (see Appendix 5.1) for any smooth convex functions whose optima form a bounded set. The one-level algorithm can also be used as the disaggregation step (5.1.11) in the two-level algorithm.

5.3 ITERATIVE AGGREGATION ALGORITHMS BASED ON CONJUGATE GRADIENT METHODS

The iterative aggregation algorithms examined so far generalize classical optimization methods by extending the one-dimensional optimization to a J-dimensional optimization. This is achieved by partitioning the vector that defines the direction of the one-dimensional optimization into the direct sum of J coordinate projections. The search takes place in a subspace generated by the projections; that is, the one-dimensional optimization "grows up" to a J-dimensional optimization.

In this section we present another technique that replaces Q of the one-dimensional optimizations with one Q-dimensional optimization. Such an approach can be realized in the method of conjugate gradients and we call the resulting algorithm the method of conjugate subspaces. Among the classical methods which have a rate of convergence faster than linear, many variations of the conjugate gradient method have the desirable feature that they do not require storing the Hessian matrix or its inverse. This feature of conjugate gradient algorithms is critical when solving problems of high dimensions. Therefore, in this book we study only iterative aggregation algorithms based on conjugate gradient methods, even though the results would apply equally as well to other conjugate direction algorithms.

The main idea underlying the method of conjugate subspaces, as well as most conjugate direction algorithms, is the minimization of some given quadratic function. When applied to nonquadratic but sufficiently smooth functions, the methods are based on the fact that close to a minimum point the original functions are well approximated by a quadratic function. We will first give an informal description of using the method of conjugate subspaces to solve (5.1.1), (5.1.2) when the function to be minimized is quadratic. In matrix notation the problem is

$$f(x) = (c,x) - \frac{1}{2}(x,Hx), \tag{5.3.1}$$

where $c, x \in \mathbb{R}^K$ and H is a symmetric positive-definite $K \times K$ matrix. If we transform the variables according to

$$x = Pz, \tag{5.3.2}$$

where P is a $K \times K$ matrix, the problem becomes

$$f(x) = f(Pz) = (c,Pz) - \frac{1}{2}(z,P'HPz). \tag{5.3.3}$$

Suppose that the columns of the $K \times K$ matrix P as well as the corresponding components of z_k are partitioned into N nonempty disjoint subsets (this partitioning is not part of the algorithm but is used to simplify the explanation of the algorithm). Without loss of generality we assume that the columns of each subset are consecutively ordered in P. Hence

$$P = (P_1, P_2, \ldots, P_N), \qquad z = (z^1, z^2, \ldots, z^N), \tag{5.3.4}$$

where P_σ and z^σ are the σth submatrix and subvector, respectively, corresponding to the partition described above. Since P is a nonsingular matrix each submatrix P_σ is of full rank with the rank equal to J^σ, the number of columns in the submatrix. Moreover, each P_σ determines a J^σ-dimensional subspace in $\mathbb{R}^K$ generated by its column vectors. We denote this subspace by $\hat{P}_\sigma$.

Two vectors a, b $\in \mathbb{R}^K$ are said to be conjugate with respect to H if (a,Hb) = 0. Two subspaces of $\mathbb{R}^K$ are said to be conjugate if any vector from one space is conjugate to every vector in the other space. The condition

$$P'_\sigma H P_\tau = 0 \qquad (\sigma \neq \tau) \tag{5.3.5}$$

is a necessary and sufficient condition for $\hat{P}_\sigma$ and $\hat{P}_\tau$ to be conjugate subspaces.

If (5.3.5) holds for all σ, τ, the matrix P'HP will be block diagonal:

$$P'HP = \begin{pmatrix} P'_1HP_1 & & & & 0 \\ & \cdot & & & \\ & & \cdot & & \\ & 0 & & \cdot & \\ & & & & P'_NHP_N \end{pmatrix} \tag{5.3.6}$$

and (5.3.3) becomes

$$f(x) = \bar{f}(z^1, \ldots, z^N) = \sum_{\sigma=1}^{N} \left[(c, P_\sigma z^\sigma) - \frac{1}{2} (z^\sigma, P'_\sigma H P_\sigma z^\sigma) \right].$$

Thus the unconstrained optimization problem (5.3.1) decomposes into N independent problems where problem σ has the form

$$\max \left\{ (c, P_\sigma z^\sigma) - \frac{1}{2} (z^\sigma, P'_\sigma H P_\sigma z^\sigma) \right\}, \qquad z^\sigma \in \mathbb{R}^{J^\sigma}, \tag{5.3.7}$$

and the vector x is determined according to (5.3.2):

$$x = \sum_{\sigma=1}^{N} P_\sigma z^\sigma. \tag{5.3.8}$$

In the algorithm to be described below, the matrices P_σ are not known in advance but rather are computed sequentially starting with an arbitrary matrix P_1. Given the matrix P_σ, let z^σ denote the solution to (5.3.7). As in the method of conjugate gradients to obtain the matrix $P_{\sigma+1}$, we need one auxiliary matrix G_σ of the same dimensions as P_σ (in conjugate gradient methods G_σ is the gradient evaluated at a given point). The sum (5.3.8) is updated after we compute P_σ and z^σ:

$$x^\sigma = x^{\sigma-1} + P_\sigma z^\sigma, \qquad \sigma = 1, \ldots, N - 1. \tag{5.3.9}$$

Equation (5.3.9) is the matrix form of (5.1.9). Thus the method of conjugate subspaces belongs to those algorithms that are based on general linear aggregation.

Each of the specific algorithms that can be described by the general form (5.3.7), (5.3.9) are determined by the choice of P_σ. In our algorithm, follow Shpalensky (1981). Let G_σ ($\sigma = 1, \ldots, N - 1$) denote a matrix "analog" of the gradient at a point x^σ, where the matrix $G_{\sigma-1}$ has the same dimensions as the matrix P_σ. We shall assume temporarily that all these matrices are $K \times J$. Let $Y_\sigma = G_\sigma - G_{\sigma-1}$. Then the analog of the conjugate gradient algorithm is

$$P_1 = G_0; \tag{5.3.10}$$

$$P_{\sigma+1} = G_\sigma - P_\sigma(Y'_\sigma P_\sigma)^{-1} Y'_\sigma G_\sigma, \tag{5.3.11}$$

or alternately, instead of (5.3.11),

$$P_{\sigma+1} = G_\sigma + P_\sigma(G'_{\sigma-1} G_{\sigma-1})^{-1} G'_\sigma G_\sigma. \tag{5.3.12}$$

(See Appendix 5.2 for the mathematical theory of its multidimensional generalization.)

We will show later that for quadratic functions, (5.3.11) and (5.3.12) are the same. However, we are still left with determining the matrix analog of the gradient G_σ. The gradient at any point represents the direction of most rapid change of the function and it is not possible to replace the gradient with a matrix that has a similar meaning. Still, the proof of conjugate directions based on the conjugate gradient method uses two properties of the gradient: orthogonality of the gradient to the direction at the minimum with respect to this direction and the validity of the quasi-Newton condition. If G_σ is selected to satisfy the matrix analog of these properties, the resulting iterative aggregation algorithm will of necessity be a conjugate subspace algorithm since the proof for the one-dimensional case will carry through here with the obvious necessary changes.

Thus if the matrix G_σ is to be a matrix analog of the gradient, it should satisfy the following conditions:

$$P_\sigma' G_\sigma = 0; \tag{5.3.13}$$

$$G_\sigma - G_{\sigma-1} = HP_\sigma \alpha_\sigma, \tag{5.3.14}$$

where α_σ is a $J \times J$ matrix. The condition (5.3.14) is a matrix analog of the quasi-Newton condition (see Gill and Murray, 1974). Combining (5.3.13), (5.3.14), we obtain

$$P_\sigma' G_\sigma = P_\sigma' G_{\sigma-1} + P_\sigma' HP_\sigma \alpha_\sigma = 0.$$

The matrix $P_\sigma' HP_\sigma$ is nonsingular since P_σ is of full rank and H is positive definite. Therefore, from the last equality and (5.3.14) we obtain

$$\begin{aligned} G_\sigma &= G_{\sigma-1} - HP_\sigma (P_\sigma' HP_\sigma)^{-1} P_\sigma' G_{\sigma-1} = (I - HA_\sigma) G_{\sigma-1}, \\ A_\sigma &= P_\sigma (P_\sigma' HP_\sigma)^{-1} P_\sigma'. \end{aligned} \tag{5.3.15}$$

From (5.3.12), (5.3.13), (5.3.15) we obtain an alternative form for $P_{\sigma+1}$ for quadratic problems:

$$P_{\sigma+1} = (I - A_\sigma H) G_\sigma. \tag{5.3.16}$$

For quadratic problems the three different algorithms that use

(5.3.11), (5.3.12), and (5.3.16), respectively, turn out to be equivalent (see Appendix 5.2 for a proof of this).

There are two methods for constructing G_σ which generate two corresponding conjugate subspace algorithms.

1. Starting with some initial matrix G_0, we recompute G_σ for each iteration using (5.3.15).
2. Let G_σ be the matrix whose columns are the gradient vectors $g(x_j^\sigma) = \nabla f(x_j^\sigma)$ evaluated at the points x_j that yield the maximum of f(x) over orthogonal linear submanifolds:

$$x = x_j^{\sigma-1} + P_\sigma z, \qquad z \in \mathbb{R}^j, \qquad j = 1, \ldots, J.$$

Let us show that the second method satisfies (5.3.13), (5.3.15). For f(x) to assume a maximal value on the linear submanifold, it is necessary that the gradient vector $g(x_j^\sigma)$ be orthogonal to the manifold. For the problem under consideration this implies that $P'_\sigma g(x_j^\sigma) = 0$ $(j = 1, \ldots, J)$ must be satisfied, which is (5.3.13).

The following equality holds:

$$f(x_j^{\sigma-1} + P_\sigma z) = (c, x_j^{\sigma-1}) - \frac{1}{2}(x_j^{\sigma-1}, Hx_j^{\sigma-1}) + (c, P_\sigma z) - (x_j^{\sigma-1}, HP_\sigma z) - \frac{1}{2}(z, P'_\sigma HP_\sigma z).$$

This expression obtains its maximum with respect to z when

$$z^\sigma = (P'_\sigma HP_\sigma)^{-1}P'_\sigma(c - Hx_j^{\sigma-1}).$$

Next we have the following series of equalities:

$$\begin{aligned} g(x_j^\sigma) &= c - Hx_j^\sigma = c - H(x_j^{\sigma-1} + P_\sigma z^\sigma) \\ &= c - Hx_j^{\sigma-1} - HA_\sigma(c - Hx_j^{\sigma-1}) \\ &= (I - HA_\sigma)g(x_j^{\sigma-1}). \end{aligned}$$

Thus

$$g(x_j^\sigma) = (I - HA_\sigma)g(x_j^{\sigma-1}), \qquad j = 1, \ldots, J, \tag{5.3.17}$$

which is equivalent to equation (5.3.15).

Thus both methods satisfy (5.3.15) when the function to be maximized in (5.3.1) is quadratic. Therefore, if the original matrices G_0 coincide, both methods will generate identical sequences of matrices $\{G_\sigma\}$.

Now we give a formal description of the conjugate subspaces algorithm for finding the unconstrained maximum of an arbitrary function f(x). In Appendix 5.2 we prove that for quadratic f(x) the subspaces are indeed conjugate and therefore the algorithm if finite in this case

Two-level iterative aggregation algorithm [proposed by Shpalensky (1981)]. Assume that we are given a starting approximation x^0 and a starting matrix G_0 which satisfy the following two constraints: the matrix G_0 has a rank of J^1, and the subspace spanned by its columns contains the gradient vector $\nabla f(x^0)$. The latter constraint is satisfied if the gradient vector is one of the columns of G_0. Let $P_1 = G_0$.

Suppose that at iteration $\sigma - 1$ we have found the point $x^{\sigma-1}$ and the $K \times J^\sigma$-dimensional matrices P_σ and $G_{\sigma-1}$. Then at iteration σ we perform the following steps:

1. The center solves the aggregate problem

$$\max f(x^{\sigma-1} + P_\sigma z), \qquad z \in \mathbb{R}^\sigma. \tag{5.3.18}$$

   Let z^σ denote the solution to (5.3.18).

2. Compute x^σ as

$$x^\sigma = x^{\sigma-1} + P_\sigma z^\sigma.$$

   Let H(x) denote the $(K \times K)$-dimensional Hessian matrix of f(x). Compute $\hat{G}_\sigma$ from the formula

$$\hat{G}_\sigma = [I - H(x^\sigma)P_\sigma(P_\sigma' H(x^\sigma)P_\sigma)^{-1}P_\sigma']G_{\sigma-1}. \tag{5.3.19}$$

3. Let $J^{\sigma+1}$ denote the rank of the $(K \times J^\sigma)$-dimensional matrix $\hat{G}_\sigma$ $(J^{\sigma+1} \leq J^\sigma)$ and let G_σ be formed from any $J^{\sigma+1}$ linearly independent column vectors of the matrix $\hat{G}_\sigma$ (this can be done by

starting from the leftmost column and excluding any column that is a linear combination of the previous columns).

4. Compute the $(K \times J^{\sigma+1})$-dimensional matrix $P_{\sigma+1}$ by any one of the formulas analogous to (5.3.11), (5.3.12), or (5.3.16):

$$P_{\sigma+1} = G_\sigma - P_\sigma(Y_\sigma' P_\sigma)^{-1} Y_\sigma' G_\sigma, \tag{5.3.20}$$

where $Y_\sigma = \hat{G}_\sigma - G_{\sigma-1}$;

$$P_{\sigma+1} = G_\sigma + P_\sigma(G_{\sigma-1}' G_{\sigma-1})^{-1} \hat{G}_\sigma' G_\sigma; \tag{5.3.21}$$

$$P_{\sigma+1} = [I - A_\sigma H(x^\sigma)]G_\sigma, \tag{5.3.22}$$

where $A_\sigma = P_\sigma(P_\sigma' H(x^\sigma) P_\sigma)^{-1} P_\sigma'$.

The choice of the formula (5.3.20)-(5.3.22) generates three distinctly different algorithms. Note that if we use the conjugate direction method or quasi-Newton methods to solve (5.3.18), we need the inverse matrix $(P_\sigma' H(x^\sigma) P_\sigma)^{-1}$, which is the inverse of the Hessian matrix of the aggregate problem at the optimal point z^σ.

If the Hessian matrix $H(x^\sigma)$ is difficult to calculate, we can replace the term $H(x^\sigma)P_\sigma$ wherever it occurs with the matrix S_σ given by

$$s_i^\sigma = \nabla f(x^\sigma) - \nabla f(x^\sigma + p_i^\sigma), \qquad i = 1, \ldots, J^\sigma, \tag{5.3.23}$$

where s_i^σ and p_i^σ are the σth columns of the matrices S_σ and P_σ, respectively. For a quadratic function $f(x)$ the matrix S_σ is identical to HP_σ. The algorithm stops when $\|\nabla f(x^\sigma)\| < \varepsilon$ for some $\varepsilon > 0$.

In these algorithms the subsystems perform various auxiliary operations that involve the computation and passage of information relevant to (5.3.20)-(5.3.22). In Appendix 5.2 it is proven that if $f(x)$ is a quadratic function, the algorithm will converge in N iterations, where N is the smallest number satisfying the inequality $\Sigma_{\sigma=1}^N J^\sigma \geqq K$. However, for arbitrary $f(x)$ it is desirable to restart the algorithm with a new G_0 of the original dimensionality. Also, the algorithm can become degenerate when the matrix G_σ vanishes.

In Lemma A5.2.3 we prove that when f(x) is a quadratic function, if the subspace spanned by the columns of G_0 contains $\nabla f(x^0)$, the algorithm does not become degenerate. For arbitrary f(x) a restart must be performed if G_0 vanishes. However, even in this case it is desirable to choose G_0 to satisfy this same condition.

One-level iterative aggregation algorithms. Each subsystem j selects an initial approximation $x_j^0 \in \mathbb{R}^K$, $j = 1, \ldots, J^1$. For the matrix G_0 we use the $(K \times J^1)$-dimensional matrix which has as its columns the vectors $g(x_j^0) = \nabla f(x_j^0)$; moreover, we assume that G_0 is of rank J^1; otherwise, we eliminate dependent columns and reduce the value of J^1. We assume that $P_1 = G_0$.

Suppose that after iteration $\sigma - 1$ we have found the points $x_j^{\sigma-1}$ $(j = 1, \ldots, J)$ and the $(K \times J^\sigma)$-dimensional matrices $G_{\sigma-1}$ and P_σ. Iteration σ will then consist of the following steps:

1. Subsystem j solves the problem

$$f(x_j^{\sigma-1} + P_\sigma z_j) \to \max, \qquad z_j \in \mathbb{R}^J. \tag{5.3.24}$$

Let z_j^σ denote the solution to (5.3.24). Then compute

$$x_j^\sigma = x_j^{\sigma-1} + P_\sigma z_j^\sigma, \qquad g(x_j^\sigma) = \nabla f(x_j^\sigma).$$

2. The $(K \times J^{\sigma-1})$-dimensional matrix $\hat{G}_\sigma$ is formed with columns equal to $g(x_j^\sigma)$, $j = 1, \ldots, J^{\sigma-1}$. From the set of vectors $g(x_j^\sigma)$ we select the maximal linearly independent subsystem consisting of J^σ vectors $(J^\sigma \leqq J^{\sigma-1})$. Without loss of generality we assume that these vectors are the first J^σ vectors $g(x_j^\sigma)$, $j = 1, \ldots, J^\sigma$. Let G_σ be the $K \times J^\sigma$ matrix formed by them.
3. The matrix $P_{\sigma+1}$ is then calculated using any one of the formulas (5.3.20)-(5.3.22). The algorithm terminates if for some j and for $\varepsilon > 0$ it is true that $\|f(x_j^\sigma)\| < \varepsilon$. We see that in the two-level algorithm the center solves one J^σ-dimensional problem each iteration, whereas in the one-level algorithm the subsystems solve J^σ such problems each iteration.

As the final topic of this chapter we develop an optimal computational scheme for maximizing the quadratic function

$$(c,x) - \frac{1}{2}(x,Hx),$$

where H is a symmetric positive-definite matrix. As mentioned previously, for a quadratic function all the variants of the algorithm are equivalent and therefore we can use the most efficient form of (5.3.20)-(5.3.22) in order to calculate $P_{\sigma+1}$. The matrix G_σ is updated using the formula (5.3.19).

Equation (5.3.19) can be restated as

$$\hat{G}_\sigma = G_{\sigma-1} - HP_\sigma[(P_\sigma'HP_\sigma)^{-1}(P_\sigma'G_{\sigma-1})]. \tag{5.3.25}$$

For most problems it is reasonable to assume that J^σ will be much smaller than K; when this is true the most computationally costly operation will be to compute the matrix HP_σ, which requires J^σ multiplications of the $(K \times K)$-dimensional matrix H by the K-dimensional column vectors of the matrix P_σ. Solution of the entire problem would thus require $\Sigma_{\sigma=1}^N J^\sigma$ such multiplications. However, it is easy to see that $K \leqq \Sigma_{\sigma=1}^N J^\sigma \leqq K + J^1 - 1$. If we assume that J^1 is considerably smaller than K, the matrix H will be multiplied by approximately K vectors. Hence in this method the number of such multiplications is roughly the same as that required for the conjugate gradient method.

Let us now examine the most effective way to perform the computations. Since some linear combination of the columns of G_0 must be equal to the gradient vector $g(x^0)$ evaluated at x^0, and since if the columns of G are linearly dependent, we exclude these columns with larger numbers, it is reasonable to assume that the first column of G_0 is equal to $g(x^0)$. When this is the case it is proven (see Theorem A5.2.3) that the first column of G_σ will be the vector $g(x^\sigma)$ for all σ. Therefore, x^σ can be determined from

$$x^\sigma = x^{\sigma-1} + A_\sigma g^{\sigma-1},$$

where $g^{\sigma-1}$ is the first column of $G_{\sigma-1}$. Using this formula, we have

$$\begin{aligned} x^{\sigma} &= x^{\sigma-1} + P_{\sigma}(P_{\sigma}'HP_{\sigma})^{-1}P_{\sigma}'g^{\sigma-1} \\ &= x^{\sigma-1} + P_{\sigma}[(P_{\sigma}'HP_{\sigma})^{-1}G_{\sigma-1}'g^{\sigma-1}] \\ &= x^{\sigma-1} + P_{\sigma}[(P_{\sigma}'HP_{\sigma})^{-1}]d^{\sigma} \\ &= x^{\sigma-1} + P_{\sigma}s^{\sigma}, \end{aligned} \tag{5.3.26}$$

where d^{σ} is the first column of the matrix $G_{\sigma-1}'G_{\sigma-1}$ and s^{σ} is the first column of the matrix $(P_{\sigma}'HP_{\sigma})^{-1}(G_{\sigma-1}'G_{\sigma-1})$.

The algorithm requires the computation of the two symmetric matrices $(P_{\sigma}'HP_{\sigma})$ and $G_{\sigma-1}'G_{\sigma-1}$. This requires performing $[J^{\sigma}(J^{\sigma} + 1)]/2$ and $[J^{\sigma-1}(J^{\sigma-1} + 1)]/2$ inner products between K-dimensional vectors. The easiest way to obtain $V^{\sigma} = (P_{\sigma}'HP_{\sigma})^{-1}G_{\sigma-1}'G_{\sigma-1}$ is to solve for v^{σ} the system of equations

$$(P_{\sigma}'HP_{\sigma})V^{\sigma} = G_{\sigma-1}'G_{\sigma-1}. \tag{5.3.27}$$

When (5.3.27) is solved using Jordan-Gaussian elimination, the number of arithmetic operations required is not more than twice the number required to solve the simple system

$$(P_{\sigma}'HP_{\sigma})y = b$$

for some vector b with vector y unknown.

Having determined V^{σ}, the matrix G_{σ} is obtained by multiplying the $(K \times J^{\sigma})$-dimensional matrix HP_{σ} by the $(J^{\sigma} \times J^{\sigma})$-dimensional vector v^{σ} and then adding two $(K \times J^{\sigma})$-dimensional matrices; x^{σ} can be found by multiplying the $(K \times J^{\sigma})$-dimensional matrix P_{σ} by the K-dimensional vector and then adding two K-dimensional vectors.

One iteration of the algorithm is completed by calculating the matrix $P_{\sigma+1}$. This can be accomplished using any of formulas (5.3.20)-(5.3.22). Formula (5.3.20) is computationally more costly than is formula (5.3.21). To calculate (5.3.21) we first must compute the matrix

$$T_{\sigma} = (G_{\sigma-1}'G_{\sigma-1})^{-1}(\hat{G}_{\sigma}'G_{\sigma}).$$

Both of the matrices on the right-hand side of the equation have been computed previously. Thus the simplest method to find T_σ is to solve the system of equations:

$$(G'_{\sigma-1}G_{\sigma-1})T_\sigma = (\hat{G}'_\sigma G_\sigma). \tag{5.3.28}$$

The computational cost of solving this system is equal to that of solving (5.3.27).

Having determined T_σ the only other computations are to multiply P_σ and T_σ and add the result to C_σ. If we compute $P_{\sigma+1}$ using (5.3.22),

$$P_{\sigma+1} = G_\sigma - P_\sigma[(P'_\sigma HP_\sigma)^{-1}P'_\sigma HG_\sigma];$$

that is, we determine T_σ by the formula

$$T_\sigma = (P'_\sigma HP_\sigma)^{-1}(P'_\sigma HG_\sigma)$$

by solving the system

$$(P'_\sigma HP_\sigma)T_\sigma = P'_\sigma HG_\sigma. \tag{5.3.29}$$

This would require computing $P'_\sigma H$ and G_σ. But (5.3.29) has the same coefficient at the unknown matrix as does the system (5.3.27), which has already been solved. Therefore, if J^σ is large, it is better to compute $P_{\sigma+1}$ from (5.3.22). However, J^σ is usually much smaller than K; in this case solving (5.3.28) would be easier than computing $P'_\sigma HG_\sigma$. Therefore, if $J^\sigma \ll K$, it is more efficient to calculate $P_{\sigma+1}$ from (5.3.21).

APPENDIX 5.1 CONVERGENCE OF ITERATIVE AGGREGATION ALGORITHMS FOR UNCONSTRAINED OPTIMIZATION PROBLEMS

In this appendix we examine iterative aggregation algorithms for solving the problem

$$f(x) \to \max; \qquad x \in \mathbf{R}^G. \tag{A5.1.1}$$

To simplify the discussion we assume throughout that f(x) is a smooth, concave function with a bounded set of maxima. According to Fiacco and McCormick (1968, Th. 24), any level set of the form $f(x) \leqq c$ for c some arbitrary constant is bounded under these assumptions.

The convergence of both the one- and two-level algorithms, except for the conjugate subspace algorithm, follows from Polak (1971, Chap. 1, Ths. 3 and 10). Moreover, it follows from the same two theorems that for an arbitrary f(x) the gradient will be zero at any limit point of these algorithms.

Polak's theorems are similar to theorems that use Lyapunov functions to prove convergence and Polak uses some function c(x) which he introduced as some kind of Lyapunov function. We shall call a point "preferable" if it is either a solution point of the original problem or worthy of special examination, such as the saddle points of the function f(x). All other points will be said to be "nonpreferred." We now present the most general of Polak's theorems.

THEOREM A5.1.1 (Polak, 1971, Th. 10) Assume that there is a process described by

$$A:\quad T \to 2^T, \qquad c:\quad T \to \mathbb{R}^1,$$

where T is some bounded subset of the Banach space B, 2^T is the set of all subsets of T, A is a mapping of T into 2^T, $\mathbb{R}^1$ is the real line, and this process consists of the following steps:

0. Take the initial point $z^0 \in T$.
1. Let $\sigma = 1$.
2. Compute $y = A(z^{\sigma-1})$.
3. Let $z^\sigma = y$.
4. If $c(z^\sigma) \leqq c(z^{\sigma-1})$, stop; otherwise, set $\sigma = \sigma + 1$ and go to step 2.

Suppose that (a) c(z) is either continuous at all nonpreferred points $z \in T$ or is bounded from below for all $z \in T$; (b) for each nonpreferred point $z \in T$ there exists numbers $\varepsilon(z) > 0$ and $\delta(z) > 0$ such that

$$c(z') - c(z'') \geqq \delta(z) > 0$$

for all $z' \in T$ for which $\|z' - z\|_B \leqq \varepsilon(z)$ and all $z'' \in A(z')$. (Here $\|\cdot\|_B$ denotes the norm of the space B.) Then the sequence (z^σ) is either finite and its next-to-last element is a preferred point or it is infinite and each of its limit points is a preferable point.

In the iterative aggregation algorithms of this chapter we consider a point to be preferable if the gradient of f(x) is zero at that point. The following theorem follows from Polak's theorems.

THEOREM A5.1.2 Given that the assumptions of Theorem A5.1.1 are valid for the algorithm A: $\mathbb{R}^G \to 2^{\mathbb{R}^G}$. If B is an algorithm satisfying $f(A(x)) \leqq f(B(x))$ for all x, any limit point of B is either a local maximum or a saddle point of f(x).

The convergence of the iterative aggregation algorithms that use a multidimensional search procedure and which generalize algorithms which use optimal line searches follows from Theorem A5.1.2. Let A be a convergent algorithm with one-dimensional optimization:

$$A(x^{\sigma-1}) = x^\sigma,$$

$$f(x^\sigma) = \max_\alpha f(x^{\sigma-1} + \alpha \Delta x^{\sigma-1}),$$

such that $\Delta x^{\sigma-1}$ depends only on $x^{\sigma-1}$. Let algorithm B be a multidimensional optimization algorithm:

$$B(\tilde{x}^{\sigma-1}) = \tilde{x}^\sigma,$$

$$f(\tilde{x}^\sigma) = \max_{\alpha_i, i=1,\ldots,N} f\left(\tilde{x}^{\sigma-1} + \sum_{i=1}^{N} \alpha_i \widetilde{\Delta x}_i^{\sigma-1}\right),$$

and again vectors $\widetilde{\Delta x}_o^{\sigma-1}$ depend only on $\tilde{x}^{\sigma-1}$. Then Theorem A5.1.2 yields the convergence algorithm B on condition that vector $\Delta x^{\sigma-1}$ lies in the subspace generated by vectors $\widetilde{\Delta x}_i^{\sigma-1}$ when $x^{\sigma-1} = \tilde{x}^{\sigma-1}$. As an example of the applications of these methods, take A to be the gradient method and B to be algorithm (5.1.13), (5.1.16)-(5.1.19). Thus we get the following:

COROLLARY Algorithm (5.1.13), (5.1.16)-(5.1.19) converges whenever the gradient method converges.

In conclusion we should point out that this theorem also applies to the conjugate subspace algorithm if after no more than a fixed number of steps N we perform a restart and the function being maximized is strictly convex and has a bounded set of maxima.

APPENDIX 5.2 MATHEMATICAL THEORY OF CONJUGATE SUBSPACES ALGORITHM FOR THE CASE OF QUADRATIC FUNCTIONS

We assume throughout this section that the function to be maximized is a quadratic function of the form $f(x) = (c,x) - (1/2)(x,Hx)$ where H is a positive-definite symmetric matrix. An outline of our analysis is as follows. First we show that the variant (5.3.22) is a correctly defined algorithm. To do so, we show that the matrix $A_\sigma = P_\sigma(P'_\sigma HP_\sigma)^{-1}P'_\sigma$ is well defined at each iteration provided that the matrix $\hat{G}_{\sigma-1}$ is not zero (i.e., $\hat{G}_{\sigma-1}$—the matrix analog of the gradient—does not vanish). Then we prove that (5.3.20)-(5.3.22) are equivalent expressions. Next we show that for $j \neq k$ it is true that $P'_jHP_k = 0$, which is tantamount to the subspaces spanned by the columns of P_j and P_k being conjugate (hence the name of the algorithm). Since H is a positive-definite matrix, it follows that the algorithm will stop after no more than N steps, where N is defined as the smallest number satisfying the inequality $\Sigma_{\sigma=1}^{N} J^\sigma \geqq K$, where $J^\sigma \geqq 1$ (here J^σ denotes the number of columns in P_σ). Finally, we prove that the algorithm stops at iteration σ (i.e., $G_\sigma = 0$) only if x^σ is an optimal point of f(x).

We start the analysis of the conjugate subspace algorithm by examining the variant defined by (5.3.19) and (5.3.22). We prove that the inverse matrix $(P'_\sigma HP_\sigma)^{-1}$ exists.

LEMMA A5.2.1 Let $\hat{G}_\sigma \neq 0$ for some σ. Then

(a) $P'_\sigma\hat{G}_\sigma = 0.$ (A5.2.1)

(b) $P'_\sigma HP_\sigma$ is a nonsingular $(J^\sigma \times J^\sigma)$-dimensional matrix.

Proof: The proof is by induction on σ. Part (b) is true for $\sigma = 1$ since the algorithm has been constructed such that the $(K \times J^{\sigma})$-dimensional matrix P_1 is of full rank J^1 and H is positive definite. Assume that part (b) is true at iteration σ. We prove that this implies that part (a) is true at iteration σ and that part (b) is true at iteration $\sigma + 1$. From (5.3.19) we have

$$P'_{\sigma}\hat{G}_{\sigma} = P'_{\sigma}G_{\sigma-1} - P'_{\sigma}HP_{\sigma}(P'_{\sigma}HP_{\sigma})^{-1}P'_{\sigma}G_{\sigma-1} = 0,$$

which proves part (a) at iteration σ. Let $z \in \mathbb{R}^{J^{\sigma+1}}$ be a $J^{\sigma+1}$-dimensional vector satisfying $P_{\sigma+1}z = 0$. Then from (5.3.22) we obtain

$$[I - P_{\sigma}(P'_{\sigma}HP_{\sigma})^{-1}P'_{\sigma}H]G_{\sigma}z = 0.$$

If we take the inner product of this last expression with the term $G_{\sigma}z$, we obtain

$$\|G_{\sigma}z\|^2 - (P'_{\sigma}G_{\sigma}z, (P'_{\sigma}HP_{\sigma})^{-1}HG_{\sigma}z) = 0. \tag{A5.2.2}$$

It follows from (A5.2.1) that $P'_{\sigma}G_{\sigma} = 0$. This can be seen from the fact that for any matrix D with K columns the equalities $D\hat{G}_{\sigma} = 0$ and $DG_{\sigma} = 0$ are equivalent since the matrix G_{σ} by the columns of the matrix $\hat{G}_{\sigma}$ are some of the columns of $\hat{G}_{\sigma}$ and all columns of $\hat{G}_{\sigma}$ can be expressed as linear combinations of the columns of G_{σ}. (Subsequently, we will use an alternative form of this result, that is, that the equalities $D_1\hat{G}_{\sigma} = D_2\hat{G}_{\sigma}$ and $D_1G_{\sigma} = D_2G_{\sigma}$ are equivalent.)

Returning to (A5.2.2) we have $G_{\sigma}z = 0$. By construction the matrix G_{σ} has linearly independent columns; hence this implies that $z = 0$. Thus the equality $P_{\sigma+1}z = 0$ implies that $z = 0$ and hence that the $(K \times J^{\sigma+1})$-dimensional matrix $P_{\sigma+1}$ is of full rank $J^{\sigma+1}$ and that the inverse matrix $(P'_{\sigma+1}HP_{\sigma+1})^{-1}$ exists because H is positive definite. This completes the proof.

The following lemma proves various matrix equalities that we need later in this section.

LEMMA A5.2.2 The following equalities hold:

(a) $P'_{\sigma-1}HP_{\sigma} = 0.$ (A5.2.3)

(b) $G'_{\sigma-1}P_{\sigma} = G'_{\sigma-1}G_{\sigma-1}.$ (A5.2.4)

(c) $A_{\sigma}HG_{\sigma-1} = P_{\sigma}.$ (A5.2.5)

(d) $P'_{\sigma-1}G_{\sigma} = 0.$ (A5.2.6)

(e) $G'_{\sigma-1}G_{\sigma} = 0.$ (A5.2.7)

Proof: (a) Substitution of the equality (5.3.22) for P in (A5.2.3) yields

$$P'_{\sigma-1}HG_{\sigma-1} - P'_{\sigma-1}HP_{\sigma-1}(P'_{\sigma-1}HP_{\sigma-1})^{-1}P'_{\sigma-1}HG_{\sigma-1} = 0.$$

(b) If we substitute the expression (5.3.22) for P_{σ} in (A5.2.4), we obtain

$$G'_{\sigma-1}P_{\sigma} = G'_{\sigma-1}G_{\sigma-1} - G'_{\sigma-1}P_{\sigma-1}(P'_{\sigma-1}HP_{\sigma-1})^{-1}P'_{\sigma-1}HG_{\sigma-1}$$
$$= G'_{\sigma-1}G_{\sigma-1}.$$

In the last equality we used the transpose of (A5.2.1) for index $\sigma - 1$:

$$\hat{G}'_{\sigma-1}P_{\sigma-1} = 0.$$

(c) Substituting for $G_{\sigma-1}$ on the left-hand side the expression $P_{\sigma} + A_{\sigma-1}HG_{\sigma-1}$ from (5.3.22) and calculating the matrices A_{σ}, $A_{\sigma-1}$ from their definitions, we obtain

$$A_{\sigma}HG_{\sigma-1} = P_{\sigma}(P'_{\sigma}HP_{\sigma})^{-1}P'_{\sigma}H(P_{\sigma} + P_{\sigma-1}(P'_{\sigma-1}HP_{\sigma-1})^{-1}P'_{\sigma-1}HG_{\sigma-1}).$$

Removing the parentheses and using (A5.2.3) yields that the last expression is equal to P_{σ}.

(d) We need to prove first that $P'_{\sigma-1}\hat{G}_{\sigma} = 0$. To that end we replace $\hat{G}_{\sigma}$ with the right-hand side of (5.3.19):

$$P'_{\sigma-1}\hat{G}_{\sigma} = P'_{\sigma-1}G_{\sigma-1} - P'_{\sigma-1}HP_{\sigma}(P'_{\sigma}HP_{\sigma})^{-1}P'_{\sigma}G_{\sigma-1} = 0.$$

From (A5.2.1) the first term is zero and the second term is also zero from (A5.2.3); hence it follows that (A5.2.6) is valid.

(e) Due to (A5.2.1) we know that $P'_\sigma \hat{G}_\sigma = 0$, as we have noted above; this is equivalent to $P'_\sigma G_\sigma = 0$. Thus substituting P_σ from (5.3.22), we get

$$0 = P'_\sigma G_\sigma = G'_{\sigma-1}G_\sigma - G'_{\sigma-1}HP_{\sigma-1}(P'_{\sigma-1}HP_{\sigma-1})^{-1}P'_{\sigma-1}G_\sigma;$$

The second term is zero due to (A5.2.6). This then implies that $G'_{\sigma-1}G_\sigma = 0$. This completes the proof of the lemma.

THEOREM A5.2.1 The equations (5.3.20)-(5.3.22) are equivalent.

Proof: To prove the equivalence of (5.3.21), (5.3.22) it is sufficient to prove that

$$P_\sigma(G'_{\sigma-1}G_{\sigma-1})^{-1}\hat{G}'_\sigma G_\sigma = -A_\sigma HG_\sigma. \tag{A5.2.8}$$

We will prove that the following equivalent equality is valid:

$$P_\sigma(G'_{\sigma-1}G_{\sigma-1})^{-1}\hat{G}'_\sigma\hat{G}_\sigma = -A_\sigma H\hat{G}_\sigma. \tag{A5.2.9}$$

Substituting the right-hand side of (5.3.19) for G_σ in the left-hand side of (A5.2.9) yields the following expression for the left-hand side:

$$P_\sigma(G'_{\sigma-1}G_{\sigma-1})^{-1}G'_{\sigma-1}(I - A_\sigma H)(I - HA_\sigma)G_{\sigma-1}. \tag{A5.2.10}$$

We have used the fact that both A_σ and H are symmetric matrices. Expanding terms and using the equalities

$$G'_{\sigma-1}A_\sigma HG_{\sigma-1} = G'_{\sigma-1}HA_\sigma G_{\sigma-1} = G'_{\sigma-1}G_{\sigma-1}$$

[from (A5.2.4), (A5.2.5)] and $HA_\sigma G_{\sigma-1} = G_{\sigma-1} - \hat{G}_\sigma$, we have that (A5.2.10) is equal to

$$P_\sigma(G'_{\sigma-1}G_{\sigma-1})^{-1}[-G'_{\sigma-1}G_{\sigma-1} + G'_{\sigma-1}P_\sigma(P'_\sigma HP_\sigma)^{-1}P_\sigma H(G_{\sigma-1} - \hat{G}_\sigma)].$$

Applying (A5.2.4) to the second term in brackets, we rewrite the same expressions:

$$P_\sigma[-I + (P'_\sigma HP_\sigma)^{-1}P'_\sigma H(G_{\sigma-1} - \hat{G}_\sigma)] = -P_\sigma + A_\sigma HG_{\sigma-1} - A_\sigma H\hat{G}_\sigma$$
$$= -A_\sigma H\hat{G}_\sigma,$$

where we have used (A5.2.5) to obtain the result. Thus (A5.2.9) is valid, which implies (A5.2.8) and thus (5.3.21) and (5.3.22) are equivalent expressions. This completes the proof of the theorem.

We now prove that the conjugate subspaces algorithm converges for quadratic functions after N iterations, where N satisfies $\Sigma_{\sigma=1}^{N} J^\sigma \geqq K$. To prove this it is sufficient to show that the J^σ-dimensional subspaces $P_\sigma \subset \mathbb{R}^K$ spanned by the columns of the matrix P_σ are conjugate subspaces. This result is proven in the following theorem.

THEOREM A5.2.2 The following equality holds:

$$P'_j HP_k = 0 \quad \text{for } j \neq k; \qquad j, k = 1, 2, \ldots, \sigma \leqq N. \tag{A5.2.11}$$

Proof: Equation (A5.2.11) is equivalent to the relation $P'_k HP_j = 0$ since $(P'_j HP_k)' = P'_k HP_j$. Therefore, we can restrict our attention to the case $j > k$. We proceed by showing that (A5.2.11) is equivalent to the relation

$$P'_j Y_k = 0. \tag{A5.2.12}$$

Substituting the expanded definition of Y_k into (A5.2.12) yields

$$P'_j Y_k = P'_j(\hat{G}_k - G_{k-1}) = P'_j(G_{k-1} - HA_k G_{k-1} - G_{k-1})$$
$$= -P'_j HP_k(P'_k HP_i)^{-1}P'_k G_{k-1}$$
$$= -(P'_j HP_k)(P'_k HP_k)^{-1}(G'_{k-1}G_{k-1})$$

The last equality is obtained using (A5.2.4). Since the last two terms are nonsingular, it follows that (A5.2.11) and (A5.2.12) are equivalent. Thus instead of proving (A5.2.11), we will prove that

$$P'_j Y_k = 0 \quad \text{for } j > k, \qquad j, k = 1, 2, \ldots, \sigma \leqq N. \tag{A5.2.13}$$

The proof is by induction on σ, where $j, k \in \{1,\ldots,\sigma\}$, $\sigma = 2, 3, \ldots$.

For $\sigma = 2$ the equality (A5.2.11) and consequently (A5.2.13) follow directly from (A5.2.3). Assume that (A5.2.13) is valid for $j, k \in \{1, \ldots, \sigma - 1\}$ with $j > k$. To show that the relation is true at σ it is sufficient to show that

$$P_\sigma' Y_k = 0 \qquad \text{for } k < \sigma - 1$$

[for $k = \sigma - 1$ the proof follows from (A5.2.3)]. The proof consists of three steps.

1. The equality

$$P_j' G_{\sigma-1} = 0 \qquad \text{for } j \leqq \sigma - 1 \tag{A5.2.15}$$

follows from our inductive assumptions. In fact, the more general statement $P_j' G_k = 0$ for $j \leqq k \leqq \sigma - 1$ follows from the inductive assumptions. To see this, we proceed inductively on k. For $k = j$ this last relationship follows from (A5.2.1). Suppose that it is valid for $k \leqq \sigma - 2$; the relation then holds for $k + 1$ since $0 = P_j' G_k = P_j' \hat{G}_{k+1} - P_j' Y_{k+1} = P_j' \hat{G}_{k+1}$. The last equality follows from the inductive assumption for (A5.2.13) since $j < k + 1 \leqq \sigma - 1$. Thus (A5.2.15) is proven.

2. The relation

$$G_j' G_{\sigma-1} = 0 \qquad \text{for } j < \sigma - 1 \tag{A5.2.16}$$

follows from the inductive assumptions. Using (A5.2.15) for $j + 1$ with $j \leqq \sigma - 2$, we have

$$0 = P_{j+1}' G_{\sigma-1} = G_j' G_{\sigma-1} - G_j' H P_j (P_j' H P_j)^{-1} P_j' G_{\sigma-1} = G_j' G_{\sigma-1},$$

where the second equality follows from (A5.2.15).

3. Equation (A5.2.14) is valid. Using (5.3.22) we have

$$P_\sigma' Y_k = G_{\sigma-1}' Y_k - G_{\sigma-1}' H P_{\sigma-1} (P_{\sigma-1}' H P_{\sigma-1})^{-1} [P_{\sigma-1} Y_k].$$

Since the expression in brackets is equal to zero by the inductive argument it follows that

$$P_\sigma' Y_k = G_{\sigma-1}' Y_k = G_{\sigma-1}' \hat{G}_k - G_{\sigma-1}' G_{k-1}.$$

Since $k < \sigma - 1$ it follows from (A5.2.16) that both terms in the last expression are zero.

This completes the proof of Theorem A5.2.2.

Theorem A5.2.2 implies that the algorithm will stop after N iterations. But the algorithm has been constructed in a manner that it will stop only when the matrix $\hat{G}_\sigma$ is identically zero for $\sigma = N$. We now prove that if $\hat{G}_\sigma = 0$, then x^σ is the maximal point of f(x).

THEOREM A5.2.3 If the $(K \times J^1)$-dimensional matrix G_0 is chosen such that the subspace spanned by its columns contains the vector $\nabla f(x^0)$ and if $\hat{G}_\sigma = 0$, then x^σ is a point where f(x) achieves a maximum.

Proof: It is sufficient to prove that the subspace spanned by the columns of the matrix $\hat{G}_\sigma$, denoted by Γ_σ, contains $\nabla f(x^\sigma)$ when $\sigma \geqq 0$. If this is true, then since $\hat{G}_\sigma = 0$ it follows immediately that $\nabla f(x^\sigma) = 0$. The proof is by induction. For $\sigma = 0$ this is true by assumption on the matrix G_0.

Assume that $\nabla f(x^{\sigma-1}) \in \Gamma_{\sigma-1}$. In matrix notation this is equivalent to the statement

$$\nabla f(x^{\sigma-1}) = G_{\sigma-1} y^{\sigma-1} \tag{A5.2.17}$$

for some vector $y^{\sigma-1} \in \mathbf{R}^J$. From (5.3.17) we have

$$\nabla f(x^\sigma) = (I - HA_\sigma)\nabla f(x^{\sigma-1}).$$

From (A5.2.17) and (5.3.19) this yields

$$\nabla f(x^\sigma) = \hat{G}_\sigma y^{\sigma-1}. \tag{A5.2.18}$$

This equality implies that $\nabla f(x^\sigma) \in \Gamma_\sigma$ and completes the proof of the theorem.

For the one-level conjugate subspace algorithm it is important to consider what situation would correspond to the condition $J^{\sigma+1} < J^\sigma$. Recall that in this algorithm column j of G_σ (let us denote it by g_j^σ) is the gradient of our function at the point x_j^σ.

LEMMA A5.2.3 Assume that the vectors $g_1, \ldots, g_{J^\sigma}$ are linearly dependent. Then either the vectors $x_1^\sigma, \ldots, x_{J^\sigma}^\sigma$ are also linearly dependent or else x*, the solution of the original problem, is a linear combination of $x_1^s, \ldots, x_{J^\sigma}^s$.

Proof: Since the vectors $g_1^\sigma, \ldots, g_{J^\sigma}^\sigma$ are linearly dependent by assumption then

$$\Sigma_i \lambda_i g_i^\sigma = 0.$$

Recall that $g_i^\sigma = -Hx_i^\sigma + c$. Hence

$$\Sigma_i \lambda_i g_i^\sigma = \Sigma_i \lambda_i(-Hx_i^\sigma + c) = -H(\Sigma_i \lambda_i x_i^\sigma) + (\Sigma_i \lambda_i)c.$$

Suppose that $\Sigma_i \lambda_i = 0$. Then $H(\Sigma_i \lambda_i x_i^\sigma) = 0$ and hence $\Sigma_i \lambda_i x_i^\sigma = 0$; that is, the x_i^σ are linearly dependent. If, however, $\Sigma_i \lambda_i \neq 0$, we can assume that $\Sigma_i \lambda_i = 1$. However, this implies that $-H(\Sigma_i \lambda_i x_i^\sigma) + c = 0$, that is, $\nabla f(x^*) = 0$, where $x^* = \Sigma_i \lambda_i x_i^\sigma$ and x* is the maximal point of f(x), which completes the proof of the lemma.

It follows from this lemma that the conjugate subspaces algorithm when used to solve quadratic problems, will rarely encounter the condition $J^{\sigma+1} < J^\sigma$, except, perhaps, at the next-to-last iteration.

6

Iterative Aggregation Algorithms for Constrained Optimization Problems

6.1 APPLICATIONS OF ITERATIVE AGGREGATION ALGORITHMS FOR UNCONSTRAINED OPTIMIZATION TO CONSTRAINED OPTIMIZATION PROBLEMS WITH VARIABLES AGGREGATION

In this chapter we turn our attention to the examination and construction of iterative aggregation algorithms for solving the problem

$$c(x) \to \max; \tag{6.1.1}$$

$$x \in Q, \tag{6.1.2}$$

where Q is a subset of $\mathbb{R}^K$.

We examine first extensions of the algorithms for unconstrained optimization, developed in Chapter 5, to solving constrained optimization problems. That is, we construct algorithms that are generalizations of relaxation methods with optimal line searches for solving constrained optimization problems.

First consider the next problems with block constraints:

$$c(x_1, \ldots, x_J) \to \max; \tag{6.1.3}$$

$$x_j \in Q_j, \qquad j = 1, \ldots, J, \tag{6.1.4}$$

where x_j is a $|K_j|$-dimensional vector and Q_j is a closed convex set in $\mathbb{R}^{|K_j|}$. The algorithms developed in Sections 5.1 and 5.2 can be used to solve this problem if the following changes are made. First, we require that the initial approximation vectors x_j^0 lie in Q_j, j = 1, ..., J. Second, the VDIs should exist for each $x_j^{\sigma-1}$ and they should not leave the set Q_j. Therefore, if $x_j^{\sigma-1}$ is an interior point of Q_j, $\Delta x_j^{\sigma-1}$ can be any vector. If, however, $x_j^{\sigma-1}$ lies on the boundary of Q_j, there must exist numbers $\bar{\alpha}_j^{\sigma} > \underline{\alpha}_j^{\sigma}$ such that the vector $x_j^{\sigma-1} + \alpha\Delta x_j^{\sigma-1}$ lies in Q_j for all α, $\underline{\alpha}_j^{\sigma} \leqq \alpha \leqq \bar{\alpha}^{\sigma}$. If for some vector $\widetilde{\Delta x}_j^{\sigma-1}$ this condition does not hold (i.e., if $x_j^{\sigma-1} + \alpha\widetilde{\Delta x}_j^{\sigma-1} \notin Q_j$ for all $\alpha \neq 0$), then we can take $\Delta x_j^{\sigma-1}$ as the solution vector v of the problem

$$\max(\widetilde{\Delta x}_j^{\sigma-1}, v); \tag{6.1.5}$$

$$x^{\sigma-1} + v \in Q_j, \qquad \|v\| \leqq \mu, \tag{6.1.6}$$

where $\mu > 0$ is a number specified in advance and v is unknown.

Also, the solution z_j^{σ} (j = 1, ..., J) of the aggregate problem must be found in a manner such that the corresponding points x_j^{σ} do not leave the sets Q_j. To achieve this, we construct the aggregate problem of the form

$$\hat{c}(z_1, \ldots, z_J) \to \min; \tag{6.1.7}$$

$$\underline{\alpha}_j^{\sigma} \leqq z_j \leqq \bar{\alpha}_j^{\sigma}, \qquad j = 1, \ldots, J. \tag{6.1.8}$$

The numbers $\underline{\alpha}_j^{\sigma}$ and $\bar{\alpha}_j^{\sigma}$ are chosen such that $x_j^{\sigma-1} + \underline{\alpha}_j^{\sigma}\Delta x_j^{\sigma-1}$ and $x_j^{\sigma-1} + \bar{\alpha}_j^{\sigma}\Delta x_j^{\sigma-1}$ are on the boundary of Q_j. Our algorithm is constructed in a manner such that for each j there is only one VDI. This is because the constraints (6.1.4) lead to simple constraints (6.1.8) in the aggregate problem. For the case where for some j there exist several VDIs,

$$\Delta x_j^{1,\sigma-1}, \ldots, \Delta x_j^{R_j,\sigma-1},$$

the constraints (6.1.8) in the aggregate problem have to be replaced by the considerably more complex constraints:

$$x_j^{\sigma-1} + \sum_{r=1}^{R_j} z_j^r \Delta x_j^{r,\sigma-1} \in Q_j.$$

(Here there are $\Sigma_{j=1}^{J} R_j$ variables z_j^r, $j = 1, \ldots, J$, $r = 1, \ldots, R_j$ in the aggregate problem.) For example, if Q_j coincides with $\mathbb{R}_+^{|G_j|}$, which is generally the case, instead of (6.1.8) there will be $G = \Sigma_j |G_j|$ constraints of the form

$$x_g^{\sigma-1} + \sum_{r=1}^{R_j} z_j^{r,} \Delta x_g^{r,\sigma-1} \geqq 0, \qquad g \in G_j, \quad j = 1, \ldots, J$$

in the aggregate problem.

For the general case this may imply that it is more costly to solve the aggregate problem then to solve the original problem. Therefore, if the Q_j are arbitrary convex sets, we should, for each group of variables, select a VDI in which the only nonzero coordinates are in the jth group; that is, we use block aggregation of variables with each block of variables represented by only one aggregate variable.

If, however, the Q_j are parallelepipeds, a sufficient condition to ensure that the constraints (6.1.4) can be expressed in the form (6.1.8) is that the VDIs have no nonzero coordinates in common.

Using descent and projection methods we can construct related algorithms for the solution of problems with linear equalities of the form

$$\max c(x); \tag{6.1.9}$$

$$(a_i, x) = b_i, \qquad i = 1, \ldots, I, \tag{6.1.10}$$

where for each i the vector a_i has the same dimension as the vector x. In solving these problems it is necessary to ensure that the initial approximation vector x^0 satisfies the constraint (6.1.10). It is also necessary that for all σ the VDIs Δx_r^σ, $r = 1, \ldots, R^\sigma$, do not leave the set defined by the constraints (6.1.10). This means that the vectors Δx_r^σ must satisfy the equation

$$(a_i, \Delta x_r^\sigma) = 0, \qquad i = 1, \ldots, I. \tag{6.1.11}$$

If one of the existing vectors Δx_r^σ does not satisfy (6.1.11), it can be projected onto the manifold defined by these equations (this procedure is described in Chapter 1). The resulting vectors Δx_r^σ are then used as the VDIs in the algorithm. Problems of the type

$$c(x) \to \max; \tag{6.1.12}$$

$$(a_i, x) = b_i, \quad i = 1, \dots, I; \tag{6.1.13}$$

$$\underline{x}_g \leqq x_g \leqq \bar{x}_g, \quad g = 1, \dots, G, \tag{6.1.14}$$

which differ from problems (6.1.9), (6.1.10) by inequality constraints (6.1.14), can be reduced to the latter problem using standard simplex-method techniques, combined with the search on verges of parallelepipeds defined by (6.1.14).

Another way that projection techniques and aggregation can be used to solve general convex programming problems is to use well-known algorithms for solving nonlinear programming problems that reduce the original problem to that of solving a sequence of minimization problems on a set of simple structure. Such methods include penalty function methods and modified Lagrangian algorithms.

Although the constraints in these problems are only implicitly present in the functionals, we cannot conclude that the constraints are being aggregated since aggregation does not imply changes of the physical meaning. Therefore, we describe such algorithms as one-sided since they only aggregate the variables of the original problem.

6.2 ONE-SIDED ITERATIVE AGGREGATION ALGORITHMS FOR SOLVING CONVEX PROGRAMMING PROBLEMS WITH CONSTRAINTS AGGREGATION

In constrained minimization problems both the variables and the constraints can be aggregated. In Section 6.1 we discussed algorithms that aggregate variables but not constraints. Thus before examining algorithms that use both variable and constraint aggregation, it seems only natural to discuss algorithms where only the constraints are aggregated. In addition to the intrinsic importance of such algorithms, they will help to develop our understanding of the process involved in aggregating constraints for the algorithms to be described later that use both constraint and variable aggregation to solve the programming problem.

Before we present these algorithms we must define more precisely what we mean by constraint aggregation. In this section we will always assume that the problem to be solved has the form

$$f_s(\xi_1,\ldots,\xi_K) = 0, \qquad s \in S; \tag{6.2.1}$$

$$x = (\xi_1,\ldots,\xi_K) \in Q; \tag{6.2.2}$$

$$c(\xi_1,\ldots,\xi_K) \to \max; \tag{6.2.3}$$

that is, we have restricted our attention to equality constraints. Problems with inequality constraints can be converted to this form by the use of supplemental variables. Theoretically, constraint aggregation consists of replacing (6.2.1) with other (aggregate) constraints. In the upper-level problem an aggregate constraint can represent several constraints from the original problem. But since the original constraints often are fundamentally disparate (e.g., from the economic standpoint one constraint may be fundamentally limiting, whereas another constraint is relatively unimportant), the relative weight given to each constraint in forming the aggregate constraint should in some way reflect this disparity. It seems clear that this relative weight should reflect the economic importance of a detailed constraint in comparison to the other constraints forming the aggregate constraint. In turn, the economic importance of a constraint is measured by its shadow price. Therefore, to aggregate constraints correctly, we must know the values of the appropriate dual variables. But as the final values of the dual variables are not known at the outset, the aggregation algorithm will have to iteratively adjust both the primal and dual variables. This problem can be viewed from a different perspective. Each constraint has a corresponding dual variable and the passage from the original constraints (6.2.1) to the aggregate constraints is equivalent to passing from the original dual variables to a system of aggregate dual variables. Thus constraint aggregation amounts to aggregation of the dual variables of the original problem. Let M(x,p) be a Lagrangian (classical or modified) function, where p is an $|S|$-dimensional vector of dual variables corresponding to the constraints (6.2.1).

Under fairly general conditions solution of the problem (6.2.1)-(6.2.3) can be found by finding the saddle point of the Lagrangian function M(x,p) on the set defined by (6.2.2):

$$\max_x \min_p M(x,p); \tag{6.2.4}$$

$$x \in Q. \tag{6.2.5}$$

In (6.2.4), (6.2.5) the primal variables x and the dual variables p have the same roles. Therefore, all the methods developed so far to aggregate primal variables can be used to aggregate dual variables. Using the geometric interpretation of aggregation given in Section 5.1, we see that a problem that only has constraint aggregation must be of the form

$$\min_p \max_x M(x,p); \tag{6.2.6}$$

$$x \in Q; \tag{6.2.7}$$

$$p \in T. \tag{6.2.8}$$

Different choices of the subspace T result in different methods of aggregating the constraints. Thus if T is a cone centered at the origin, we have proportional aggregation. If T is an arbitrary linear manifold, we have general linear aggregation.

We can rewrite (6.2.6)-(6.2.8) to be in a form more consistent with the concept of constraint aggregation. We will do this for the case of general linear aggregation of dual variables. Let the subspace T be defined as

$$T = \left\{p \middle| p = \left\{p_s = t_s + \sum_{i=1}^{I} \lambda_i \theta_{is} \middle| s \in S\right\}\right\}, \tag{6.2.9}$$

where $t = (t_s | s \in S)$ and $\theta_i = (\theta_{is} | s \in S)$, i = 1, ..., I, are known vectors of dimension $|S|$. If we assume that M(x,p) is the classical Lagrangian function, (6.2.6)-(6.2.8) become

$$\sum_{s\in S} \theta_{is} f_s(x) = 0, \qquad i = 1, \ldots, I; \tag{6.2.10}$$

$$x \in Q; \tag{6.2.11}$$

$$\max_x \left\{ c(x) - \sum_{s\in S} t_s f_s(x) \right\}. \tag{6.2.12}$$

If M(x,p) is the usual modified Lagrangian function, (6.2.6)-(6.2.8) become

$$\sum_{s\in S} \theta_{is} f_s(x) = 0, \qquad i = 1, \ldots, I; \tag{6.2.13}$$

$$x \in Q; \tag{6.2.14}$$

$$\max_x \left\{ c(x) - \sum_{s\in S} t_s f_s(x) - \sum_{s\in S} \frac{q}{2} f_s^2(x) \right\}. \tag{6.2.15}$$

Guidelines for choosing the most effective aggregate problem are discussed in Section 6.5. Therefore, in this section we will not be concerned with comparing the computational effectiveness of the two definitions of the Lagrangian function. Note, however, that while (6.2.13)-(6.2.15) is a more complicated problem than (6.2.10)-(6.2.12), it has a much wider scope of application.

Having given an overview of the main ideas used to construct aggregate constraints we can now turn our attention to the construction and analysis of iterative aggregation algorithms with constraint aggregation. In this section we consider only algorithms where the aggregate problem is defined by (6.2.13)-(6.2.15). These algorithms are closely related to modified Lagrangian algorithms that have been receiving much scholarly attention recently. It has proven possible to construct a general theory of iterative aggregation for these algorithms as well as to determine which algorithms are the most computationally efficient.

By analogy with the methods for forming aggregate functions in the algorithms of Chapter 5, we will assume that estimates for the following coefficients are available after iteration $\sigma - 1$:

$$t_s = p_s^{\sigma-1}, \qquad \theta_{is} = \theta_{is}^{\sigma},$$

where $p^{\sigma-1}$ is the approximation to the dual solution vector found at iteration $\sigma - 1$, and θ_i^σ is the ith VDI at iteration σ. With these assumptions, the aggregate problem (6.2.12)-(6.2.15) assumes the form

$$\sum_{s \in S} \theta_{is}^{\sigma} f_s(x) = 0, \qquad i = 1, \ldots, I; \tag{6.2.16}$$

$$x \in Q; \tag{6.2.17}$$

$$\max_x \left\{ c(x) - \sum_{s \in S} p_s^{\sigma-1} f_s(x) - \sum_{s \in S} \frac{q}{2} f_s^2(x) \right\}. \tag{6.2.18}$$

We prove in Appendix 6.1 that for all practical purposes this problem has a solution if the original problem is solvable. We also prove in the appendix the main property of (6.2.16)-(6.2.18), that the solution of this problem is equivalent to the minimization of the dual (with respect to the modified Lagrangian function) function

$$\Psi(p) = \sup_{x \in Q} \left\{ c(x) - \sum_{s \in S} p_s f_s(x) - \sum_{s \in S} \frac{q}{2} f_s^2(x) \right\}; \tag{6.2.19}$$

$$\Psi(p) \to \min,$$

on the manifold that passes through the point $p^{\sigma-1}$ and is spanned by the vectors θ_i^σ, $i = 1, \ldots, I$. The minimum of the function Ψ is obtained at the point p^σ, which in turn is equal to

$$p^\sigma = p^{\sigma-1} + \Sigma_i \lambda_i^\sigma \theta_i^\sigma, \tag{6.2.20}$$

where λ_i^σ is the dual solution of (6.2.16)-(6.2.18) and the vector $f(x^\sigma) = (f_s(x^\sigma) | s \in S)$ is equal to the antigradient of $\Psi(p)$ at p^σ: $\nabla\Psi(p^\sigma) = -f(x^\sigma)$. The solution of the original problem is equivalent to the minimization of $\Psi(p)$.

It follows from these properties of the aggregate problem that we can find the minimum of $\Psi(p)$ over an arbitrary linear manifold and compute the gradient $\Delta\Psi(\tilde{p})$ at the minimum point $\tilde{p}$. Thus even without explicitly knowing the form of $\Psi(\tilde{p})$, we can determine where it achieves a minimum. Moreover, we can use algorithms for unconstrained optimization, such as those of Chapter 5, to minimize $\Psi(p)$, and then modify the algorithms using iterative aggregation.

We now show how the existing algorithms with constraint aggregation fit into this general scheme. Martines-Soler (1977) presents a one-sided algorithm in which θ_i^σ at iteration σ has as its coordinates $\theta_{is}^\sigma = p_s^{\sigma-1}$ for $s \in S_i$, $\theta_{is}^\sigma = 0$ for $s \notin S$, and $t_s = 0$.

Note that Martines-Soler (1977) suggested the use of (6.2.15) with $t_s = 0$ only for linear programming problems; for other problems the use of c(x) is suggested as the functional for the aggregated problem. This method of choosing θ_i^σ transforms (6.2.16)-(6.2.18) into the problem

$$\sum_{s \in S} p_s^{\sigma-1} f_s(x) = 0, \qquad i = 1, \ldots, I; \tag{6.2.21}$$

$$x \in Q; \tag{6.2.22}$$

$$\max_x \left\{ c(x) - \sum_{s \in S} \frac{q}{2} f_s^2(x) \right\}, \tag{6.2.23}$$

which is the problem considered by Martines-Soler. In addition, if λ_i^σ, $i = 1, \ldots, I$, is the dual solution of (6.2.13)-(6.2.15) with given t_s, θ_{is}, and $\tilde{\lambda}_i^\sigma$, $i = 1, \ldots, I$, the dual solution of (6.2.21)-(6.2.23), then $\lambda_i^\sigma = \tilde{\lambda}_i^\sigma + 1$, $i = 1, \ldots, I$. The algorithm of Martines-Soler determines p^σ from the solution of (6.2.21)-(6.2.23) by

$$p_s^\sigma = \lambda_i^\sigma p_s^{\sigma-1} - \alpha^\sigma f_s(x^\sigma), \qquad s \in S_i; \tag{6.2.24}$$

$$\alpha^\sigma > 0, \qquad \alpha^\sigma \to 0, \qquad \Sigma_\sigma \alpha^\sigma = \infty. \tag{6.2.25}$$

Thus in this algorithm $\Psi(p)$ is minimized by means of an iterative process that has two steps per iteration. At the first step we find the minimum of $\Psi(p)$ in a cone containing $p^{\sigma-1}$, and then we move from the minimizing vector in a direction opposite to the gradient by a coefficient α^σ that tends toward zero. In Appendix 6.1 we prove that if the set of dual solutions of the original problem is bounded, the algorithm converges (Martines-Soler proves the convergence only on a subsequence).

In the algorithms of Vakhutinsky and Dudkin (1973, 1979) the dual solutions are always updated by the formula

$$p_s^\sigma = \lambda_i^\sigma p_s^{\sigma-1} - \alpha f_s(x^\sigma), \qquad s \in S_i, \tag{6.2.26}$$

where λ_i is the dual solution of the aggregate problem. Vakhutinsky and Dudkin also considered aggregate problems distinct from (6.2.16)-(6.2.18), but since the algorithm (6.2.21)-(6.2.23), (6.2.26) is a special case of algorithms they considered, we shall refer to it as the one-sided Vakhutinsky-Dudkin algorithm. We can see that in practical terms this algorithm coincides with the one-sided algorithm of Martines-Soler, differing from the latter only in the fact that the shift along the antigradient is of fixed step length. In the appendix we prove that if $\alpha < 2q$, this algorithm converges under the same conditions as that of Martines-Soler. Other things being equal, an algorithm with α^σ fixed, provided that it converges, should be computationally more efficient than an algorithm where α^σ tends toward zero. Thus the Vakhutinsky-Dudkin algorithm should have a much faster rate of convergence than the Martines-Soler algorithm.

Returning to the general scheme (6.2.16)-(6.2.18), (6.2.20) proposed for problems with constraint aggregation, note that the computational cost of (6.2.16)-(6.2.18) is almost independent of the vectors θ_i^σ. Therefore, when we turn to general linear aggregation of the dual variables, we do not increase the computational complexity of solving the aggregate problem by any practical amount. However, the improvement derived from solving this aggregate problem increases substantially since $\Psi(p)$ is being minimized over a rationally chosen manifold rather than over a fixed cone independent of its structure in the neighborhood of $p^{\sigma-1}$. More important, in updating the dual variables we do not simply take a fixed step size along the antigradient but rather move to the point that achieves the minimum in that direction. Gradient methods with full one-dimensionsl optimization converge much faster than do ones with fixed step sizes; also, and more important, they allow for the transition to more effective algorithms such as conjugate gradient algorithms. All this can be accomplished with no appreciable increase in the computational burden of the algorithm. The actual choice of λ_i^σ is determined by the actual algorithm used to minimize

$\Psi(p)$. To illustrate how an iterative aggregation algorithm can be constructed on the basis of an algorithm used to minimize $\Psi(p)$, let us analyze the algorithm that corresponds to (5.1.13), (5.1.16)-(5.1.19), that is, the algorithm that uses steepest descent with respect to the dual variables.

For the initial approximation we have the vectors p^0, θ^1 (here θ^σ is the vector constructed from nonzero components θ^σ_{is}). In this instance we use block aggregation. Given $p^{\sigma-1}$, θ^σ, after iteration $\sigma - 1$ we solve the aggregate problem

$$\sum_{s \in S_i} \theta^\sigma_s f_s(x) = 0, \qquad i = 1, \ldots, I; \tag{6.2.27}$$

$$x \in Q; \tag{6.2.28}$$

$$\max \left\{ c(x) - \sum_{s \in S} p^{\sigma-1}_s f_s(x) - \sum_{s \in S} \frac{1}{2} f^2_s(x) \right\}. \tag{6.2.29}$$

Having solved the aggregate problem, we update p^σ, $\theta^{\sigma+1}$ by

$$p^\sigma_s = p^{\sigma-1}_s + \lambda^\sigma_i \theta^\sigma_s, \qquad s \in S_i, \qquad i = 1, \ldots, I; \tag{6.2.30}$$

$$\theta^{\sigma+1}_s = f_s(x^\sigma). \tag{6.2.31}$$

When $I = 1$ we can use a conjugate gradient algorithm to minimize $\Psi(p)$. The only difference between this method and (6.2.27)-(6.2.31) lies in the fact that $\theta^{\sigma+1}_s$ is not determined by (6.2.31) but by

$$\theta^{\sigma+1}_s = -f_s(x^\sigma) + \frac{\Sigma_{t \in S} f^2_t(x^\sigma)}{\Sigma_{t \in S} f^2_t(x^{\sigma-1})} \theta^\sigma_s. \tag{6.2.32}$$

When $I \neq 1$ we can use the conjugate subspaces algorithm to minimize $\Psi(p)$ (see Section 5.3). This algorithm will also differ from (6.2.27)-(6.2.31) in the way $\theta^{\sigma+1}_i$ is updated.

Thus the essence of iterative aggregation algorithms for constraint aggregation consists of using a regular minimization algorithm to minimize $\Psi(p)$ over an arbitrary manifold. This reveals the connection with penalty methods that represent a fixed-step gradient method for minimizing $\Psi(p)$. In Appendix 6.1 we prove the global convergence of (6.2.27)-(6.2.31) assuming that $f_s(x)$ is linear for all s. When $f_s(x)$ can be nonlinear functions, it is possible to prove local convergence only if q is sufficiently large.

In conclusion, we note that in constraint aggregation we reduce the original problem to another problem with fewer constraints which can then be solved using the iterative aggregation algorithms described in Section 6.1 for variable aggregation.

6.3 TWO-LEVEL ITERATIVE AGGREGATION ALGORITHMS WITH BOTH VARIABLES AND CONSTRAINTS AGGREGATION

Before we analyze the rather complex two-level iterative aggregation algorithm that uses both variable and constraint aggregation, it is useful to give as an example a simpler algorithm that also uses constraint and variable aggregation. In this simpler algorithm variables are aggregated using linear aggregation in a manner that takes into account nonnegativity conditions.

In aggregating constraints we assume that the set S of constraints has been partitioned into I disjoint subsets S_i ($i = 1, \ldots, I$) the union of which does not necessarily coincide with S. It is also possible for this union to be the empty set ($I = 0$). We apply simple or block proportional aggregation to the dual variables from each subset $S_1, \ldots, S_I$ and fix the value of the dual variables p_s ($s \in S \backslash \cup_{i=1}^{I} S_i$) in the aggregation problem.

We assume that the original problem has global constraints as well as local constraints [which contain only variables in partition j ($j = 1, \ldots, J$)]. Denote the sets of these constraints by L_j and assume that the values of the corresponding dual variables are also fixed in the aggregate problem. Then the original problem can be written in the following decomposable format:

$$f_s(x_1,\ldots,x_J) \leqq 0, \qquad s \in S; \tag{6.3.1}$$

$$h_t(x_j) \leqq 0, \qquad t \in L_j, \qquad j = 1, \ldots, J; \tag{6.3.2}$$

$$x_j \geqq 0, \qquad j = 1, \ldots, J; \tag{6.3.3}$$

$$\max\ c(x_1,\ldots,x_J), \tag{6.3.4}$$

where x_j is a $|K_j|$-dimensional vector, S the set of global constraints, and L_j the set of local constraints for subsystem j.

We have written the problem so that the constraints involving only subsystem j are written separately. From an economics standpoint this is equivalent to saying that these constraints are for resources preallocated to the subsystem.

Suppose that after iteration $\sigma - 1$ we have obtained the approximation $x^{\sigma-1}$. To construct the aggregate problem we assume that the vector x is linearly dependent on R parameters $z_1, z_2, \ldots, z_R$ [as in (5.1.3)]:

$$x = \varphi^\sigma(z_1,\ldots,z_R), \tag{6.3.5}$$

such that $z_1 \geqq 0, \ldots, z_R \geqq 0$, $\varphi^\sigma(z_1,\ldots,z_R) \geqq 0$, $\varphi^\sigma(0,\ldots,0) = 0$. Also, for a certain nonnegative vector $e^\sigma = (e_1^\sigma,\ldots,e_R^\sigma)$ we should be able to obtain the original point, that is, $x^{\sigma-1} = \varphi^\sigma(e_1^\sigma,\ldots,e_R^\sigma)$. It is easy to see that the linear relation (6.3.5) subject to the given constraints can be expressed as

$$\xi_k = \sum_{r=1}^{R} \varphi_{kr}^\sigma z_r, \tag{6.3.6}$$

where $x = (\xi_1,\ldots,\xi_K)$, and

$$e_r^\sigma \geqq 0, \qquad \varphi_{rk}^\sigma \geqq 0, \qquad \sum_{r=1}^{R} \varphi_{kr}^\sigma e_r^\sigma = \xi_k^{\sigma-1}, \qquad k = 1, \ldots, K. \tag{6.3.7}$$

Denote $\Phi^\sigma = (\varphi_{kr}^\sigma)$ a $(K \times R)$-dimensional matrix. We assume that

$$\varphi_{kr}^\sigma = \varphi_{kr}(x^{\sigma-1}), \qquad e_r^\sigma = e_r(x^{\sigma-1}), \tag{6.3.8}$$

where $e_r(x)$, $\varphi_{kr}(x)$ are continuous functions of the K-dimensional vector x. Then (6.3.6) can be expressed in matrix form as $x = \Phi^\sigma z$, where z is the R-dimensional vector $z = (z_1,\ldots,z_R)$.

Now let us construct a two-level iterative aggregation algorithm for the problem (6.3.1)-(6.3.4). First recall that the resource set S has been partitioned into I disjoint subsets S_i, i = 1, ..., I. Also let $S_0 \subset S$ be an arbitrary subset of S (subsequently we shall construct a quadratic modified Lagrangian function for the set of resources S_0). The set K of variable indexes has been partitioned into J disjoint subsets K_j, j = 1, ..., J. Thus the vectors in (6.3.1)-(6.3.4) are $|K_j|$-dimensional, $x_j = (\xi_k | x \in K_j)$.

Let Φ_j^σ denote the ($K_j \times R$) submatrix of the ($K \times R$) matrix Φ^σ: $\Phi_j^\sigma = (\varphi_{kr}^\sigma | k \in K_j, r = 1, \ldots, R)$. Then the relation $x = \Phi^\sigma z$ can be equivalently stated as

$$(x_1, x_2, \ldots, x_J) = (\Phi_1^\sigma z, \Phi_2^\sigma z, \ldots, \Phi_J^\sigma z). \tag{6.3.9}$$

Suppose that at the end of iteration $\sigma - 1$ we have obtained the primal approximation x^σ as well as the dual approximations for the constraints (6.3.2) and (6.3.3): $p_s^{\sigma-1}$ ($s \in S$) and $p_t^{\sigma-1}$ ($t \in L_j$, j = 1, ..., J), respectively. Then iteration σ consists of the following steps:

1. Construct the aggregate and semiaggregate functions.

a. Semiaggregate resource utilization functions:

$$\hat{f}_s^\sigma(z) = f_s(\Phi^\sigma z), \qquad s \in S. \tag{6.3.10}$$

b. Aggregate resource utilization functions:

$$f_i^\sigma(z) = \sum_{s \in S_i} p_s^{\sigma-1} \hat{f}_s^\sigma(z), \qquad i = 0, \ldots, I. \tag{6.3.11}$$

c. Aggregate the objective function:

$$c^\sigma(z) = c(\Phi^\sigma z) - \sum_{j=1}^{J} \sum_{t \in L_j} p_t^{\sigma-1} h_t(\Phi_j^\sigma z). \tag{6.3.12}$$

2. Solve the aggregate problem:

$$\max \left\{ c^{\sigma}(z) - f_0^{\sigma}(z) - \sum_{s \in S_0} p_s^{\sigma-1} u_s - \frac{1}{2} Q_0' \sum_{r=1}^{R} \beta_r^{\sigma}(z_r - e_r^{\sigma})^2 \right.$$

$$\left. - \frac{1}{2} Q_0 \sum_{s \in S_0} \left[\sum_{r=1}^{R} b_{sr}^{\sigma}(z_r - e_r^{\sigma}) + \hat{f}_s^{\sigma}(e^{\sigma}) + u_s \right]^2 \right\}; \qquad (6.3.13)$$

$$f_i(z) \leqq 0, \qquad i = 1, \ldots, I; \qquad (6.3.14)$$

$$z \geqq 0, \qquad u_s \geqq 0, \qquad s \in S_0. \qquad (6.3.15)$$

Here Q_0', $Q_0 > 0$ are parameters of the solution algorithm; $\beta_r^{\sigma} > \beta > 0$, where β_r^{σ} can either be specified in advance or treated as parameters of the process or viewed as normalizing factors that continuously depend on $x^{\sigma-1}$;

$$b_{sr}^{\sigma} = \frac{\partial f_s^{\sigma}(e^{\sigma})}{\partial z_r}. \qquad (6.3.16)$$

Let z_r^{σ}, u_s^{σ} ($r = 1, \ldots, R$, $s \in S_0$) denote the primal solution of the aggregate problem λ_i^{σ} ($i = 1, \ldots, I$) the dual variables corresponding to the constraints (6.3.14). The variables $u_s \geqq 0$ are supplemental variables that transform (6.2.1) into equalities for $s \in S_0$.

3. Each subsystem $j = 1, \ldots, J$ solves a local problem. For notational convenience let

$$c_j^{\sigma}(x_j) = c(\Phi_1^{\sigma} z^{\sigma}, \ldots, \Phi_{j-1}^{\sigma} z^{\sigma}, x_j, \Phi_{j+1}^{\sigma} z^{\sigma}, \ldots, \Phi_J^{\sigma} z^{\sigma}); \qquad (6.3.17)$$

$$f_{sj}^{\sigma}(x_j) = f_s(\Phi_1^{\sigma} z^{\sigma}, \ldots, \Phi_{j-1}^{\sigma} z^{\sigma}, x_j, \Phi_{j+1}^{\sigma} z^{\sigma}, \ldots, \Phi_J^{\sigma} z^{\sigma}), \qquad (6.3.18)$$

and $g_{sj}^{\sigma}(x_j)$ denotes the $|K_j|$-dimensional gradient vector of $f_{sj}^{\sigma}(x_j)$:

$$g_{sj}^{\sigma}(x_j) = \left(\frac{\partial f_{sj}^{\sigma}(x_j)}{\partial \xi_k} \middle| k \in K_j \right). \qquad (6.3.19)$$

Using this notation the local problem can be expressed as

$$\max \left\{ c_j^\sigma(x_j) - \sum_{s \in S} \tilde{p}_s^\sigma (f_{sj}^\sigma(x_j) + v_{sj}) \right.$$

$$\left. - \frac{1}{2} Q \sum_{s \in S} [(g_{sj}^\sigma(\Phi_j^\sigma z^\sigma), x_j - \Phi_j^\sigma z^\sigma) + v_{sj}]^2 \right\}; \qquad (6.3.20)$$

$$h_t(x_j) \leqq 0, \qquad t \in L_j; \qquad (6.3.21)$$

$$x_j \geqq 0, \qquad v_{sj} \geqq 0, \qquad s \in S, \qquad (6.3.22)$$

where

$$\tilde{p}_s^\sigma = \begin{cases} \lambda_i^\sigma p_s^{\sigma-1} & s \in S_i,\ i = 1, \ldots, I; \\ p_s^{\sigma-1} & s \in S \backslash \cup_{i=1}^I S_i; \end{cases} \qquad (6.3.23)$$

$Q > 0$ is a parameter of the process. Let $\hat{x}_j^\sigma$, v_{sj}^σ denote the solution of this problem and $\hat{p}_t^\sigma$ $(t \in L_j)$ the dual variables corresponding to the constraints (6.2.21), where the v_{sj} play the role of supplemental variables. The quadratic terms in (6.3.20) have the following meaning. If we interpret (6.3.1) as representing resource constraints, we can conclude that the solution z^σ of the aggregate problem (6.3.13)-(6.3.15) says that the center plans to use $f_s(\Phi^\sigma z^\sigma)$ of resource s. If subsystem j selects plan x_j and the plans of the other subsystems that have been recommended to them by the center are fixed (i.e., $\Phi_j^\sigma z^\sigma$ is fixed for $j' \neq j$), then from (6.3.18) the expenditures of resource s will be equal to $f_{sj}^\sigma(x_j)$. It is preferable to have $f_{sj}^\sigma(x_j) \leqq f_s(\Phi^\sigma z^\sigma)$.

Using (6.3.18) we replace the right-hand side of the inequality by $f_{sj}^\sigma(\Phi_j^\sigma z^\sigma)$ and transform it into the equality

$$f_{sj}^\sigma(x_j) - f_{sj}^\sigma(\Phi_j^\sigma z^\sigma) + v_{sj} = 0, \qquad v_{sj} \geqq 0.$$

This equality can be put into the objective function as a quadratic penalty term with parameter $-Q/2$. If we linearize f_{sj} at the point $\Phi_j^\sigma z^\sigma$:

$$f_{sj}^\sigma(x_j) \sim f_{sj}^\sigma(\Phi_j^\sigma z^\sigma) + (g_{sj}^\sigma(\Phi_j^\sigma z^\sigma), x_j - \Phi_j^\sigma z^\sigma)$$

and substitute this linearization into the quadratic penalty term, we obtain the quadratic term in (6.3.20).

4. Compute the preliminary estimates of the dual variables:

$$\hat{p}_s^\sigma = (\tilde{p}_s^\sigma + \alpha Q f_s(\hat{x}_1^\sigma, \ldots, \hat{x}_J^\sigma))_{(+)}, \qquad s \in S, \tag{6.3.24}$$

where $\alpha > 0$ is a parameter of the process.

5. Compute the next approximations to the primal and dual solutions of the original problem:

$$\begin{aligned} x_j^\sigma &= (1 - \gamma)x_j^{\sigma-1} + \gamma\hat{x}_j^\sigma, & j &= 1, \ldots, J; \\ p_s^\sigma &= (1 - \gamma)p_s^{\sigma-1} + \gamma\hat{p}_s^\sigma, & s &\in S; \\ p_t^\sigma &= (1 - \gamma)p_t^{\sigma-1} + \gamma\hat{p}_t^\sigma, & t &\in L_j, \quad j = 1, \ldots, J, \end{aligned} \tag{6.3.25}$$

where γ is a parameter of the process satisfying $0 < \gamma < 1$.

In Appendix 6.2 we prove that the stationary points of the algorithm are optimal solutions of the original problem (6.3.1)-(6.3.4). We also prove the algorithm converges locally subject to the additional assumption that the local constraints (6.3.3) are either absent or treated as global constraints; we further assume that $S_0 = S$ and $I = 0$ and the optimal solution satisfies sufficient second-order conditions.

6.4 ONE-LEVEL ITERATIVE AGGREGATION ALGORITHMS WITH BOTH VARIABLES AND CONSTRAINTS AGGREGATION

In this section we present a one-level algorithm that corresponds to the two-level algorithm described in Section 6.3. As before, we differentiate between parallel and sequential algorithms. A general scheme for constructing one-level algorithms from two-level algorithms has already been used several times in this book. The main idea is based on the fact that instead of a single aggregate problem each subsystem generates its own aggregate problem, in which the variables corresponding to that subsystem are not aggregated. In this connection we have already noted that one-level algorithms are related to group coordinate descent algorithms. In addition, it turns out that for constrained optimization it is necessary to use dual variables; different techniques for including the dual variables are discussed at the end of this section.

First we analyze the parallel algorithm. Taking (6.3.1)-(6.3.4) as the original problem, we now construct the problem for subsystem j. The aggregation will be carried out by formulas analogous to (6.3.10)-(6.3.12). However, let us assume that different subsystems have different aggregating functions. In subsystem j

$$\xi_k = \xi_k, \qquad k \in K_j; \tag{6.4.1}$$

$$\xi_k = \varphi_k^{j,\sigma}(z_1^j, \ldots, z_{R_j}^j), \qquad k \notin K_j. \tag{6.4.2}$$

Moreover, if $z_1^j \geqq 0, \ldots, z_{R_j}^j \geqq 0$, the relations

$$\varphi_k^{j,\sigma}(z_1^j, \ldots, z_{R_j}^j) \geqq 0, \qquad \varphi_k^{j,\sigma}(0, \ldots, 0) = 0,$$

$$\xi_k^{\sigma-1} = \varphi_k^{j,\sigma}(1, \ldots, 1)$$

are valid and the functions $\varphi_k^{j,\sigma}$ are all linear. Using (6.4.1), (6.4.2) we aggregate the variables at iteration σ. We obtain the following functions:

1. Semiaggregate resource utilization function:

$$\hat{f}_{js}^{\sigma}(x_j, z^j) = f_s(x_j, \Phi^{j,\sigma} z^j), \tag{6.4.3}$$

   where z^j is the vector $(z_i^j \mid i = 1, \ldots, R_j, i \neq j)$ and where $\Phi^{j,\sigma}$ is analogous to the matrix Φ described in Section 6.3.

2. Aggregate resource utilization function:

$$f_{ji}^{\sigma}(x_j, z^j) = \sum_{s \in S_i} p_s^{\sigma-1} \hat{f}_{js}^{\sigma}(x_j, z^j), \qquad i = 0, 1, \ldots, I. \tag{6.4.4}$$

3. Aggregate objective function:

$$c_j^\sigma(x_j, z^j) = c(x_j, \Phi^{j,\sigma} z^j) - \sum_{\nu \neq j} \sum_{t \in L_j} p_t^{\sigma-1} h_t(\Phi_\nu^{j,\sigma} z^j), \qquad (6.4.5)$$

where $\Phi_\nu^{j,\sigma} z^j$ is the subvector of the vector $\Phi^{j,\sigma} z^j$ whose coordinate indices belong to K_ν. Here the local constraints are being considered indirectly through the functional of the problem using the Lagrangian function.

Next each subsystem j solves the problem

$$\max \left\{ c_j^\sigma(x_j, z^j) - f_{j0}^\sigma(x_j, z^j) - \sum_{s \in S_0} p_s^{\sigma-1} u_s \right.$$

$$\left. - \frac{Q_0'}{2} \sum_{r=1}^{R} \beta_r^\sigma (z_r^\sigma - 1)^2 - \frac{Q_0}{2} \sum_{s \in S_0} [\hat{f}_{js}^\sigma(x_j, z^j) + u_s]^2 \right\}; \qquad (6.4.6)$$

$$f_{ji}^\sigma(x_j, z^j) \leqq 0, \qquad i = 1, \ldots, I; \qquad (6.4.7)$$

$$h_j(x_j) \leqq 0; \qquad (6.4.8)$$

$$z^j \geqq 0, \qquad u_s \geqq 0, \qquad s \in S_0. \qquad (6.4.9)$$

In this problem we did not linearize the functions $\hat{f}_{js}^\sigma(x_j, z^j)$ which appear in the quadratic term from the expression (6.4.6). If the linearized form is preferred, it can be developed along the lines of (6.3.13)-(6.3.15).

Having solved the problem (6.4.6)-(6.4.9) we get $\hat{x}_j^\sigma$, $z^{j,\sigma}$, and the dual variables $\lambda_i^{j,\sigma}$, i = 1, ..., I, $\tilde{p}_t^\sigma$, $t \in L_j$ for the constraints (6.4.7). From these we define the vectors

$$\hat{x}^{j,\sigma} = (\hat{x}_j^\sigma, \Phi^{j,\sigma} z^{j,\sigma}); \qquad (6.4.10)$$

$$\tilde{p}_s^{j,\sigma} = \lambda_i^{j,\sigma} p_s^{\sigma-1}, \qquad s \in S_i, \qquad i = 1, \ldots, I \qquad (6.4.11)$$

(coordinates of the vector $\hat{x}^{j,\sigma}$ are those of $\hat{x}_j^\sigma$ and $\Phi^{j,\sigma} z^{j,\sigma}$ taken together). Having solved (6.4.6)-(6.4.9) for all the subsystems and having determined all the vectors above, we then calculate the vectors

$$x^\sigma = \Sigma_j \gamma_j^\sigma \hat{x}^{j,\sigma}; \tag{6.4.12}$$

$$\tilde{\tilde{p}}_s^\sigma = \Sigma_j \gamma_j^\sigma \tilde{p}_s^{j,\sigma}, \qquad s \in S_i, \qquad i = 1, \ldots, I; \tag{6.4.13}$$

$$p_t^\sigma = \gamma_j^\sigma \tilde{\tilde{p}}_t^\sigma + (1 - \gamma_j^\sigma) p_t^{\sigma-1}, \qquad t \in L_j, \qquad j = 1, \ldots, J, \tag{6.4.14}$$

where

$$\gamma_j^\sigma > \delta > 0, \qquad \Sigma_j \gamma_j^\sigma = 1. \tag{6.4.15}$$

The formula for updating the dual variables corresponding to the local constraints can be obtained from the following considerations. In this method for forming the aggregate problem, each subsystem except for subsystem j fixes the value of the dual variables of subsystem j's local constraints. Therefore, averaging over subsystems we obtain

$$p_t^\sigma = \gamma_j^\sigma \tilde{p}_t^\sigma + \sum_{i \neq j} \gamma_i^\sigma p_t^{\sigma-1}, \qquad t \in L_j, \qquad j = 1, \ldots, J,$$

and from (6.4.15) we have

$$\sum_{i \neq j} \gamma_i^\sigma = 1 - \gamma_j^\sigma,$$

and together these results yield (6.4.14).

The value of x^σ calculated from (6.4.12) is the new approximation to the primal solution at iteration σ, and $(p_t^\sigma | t \in L_j, j = 1, \ldots, J)$ is the new approximation to the vector of dual variables of the local constraints. The approximation to the dual vectors corresponding to the global constraints is determined as

$$p_s^\sigma = \{p_s^{\sigma-1} + \alpha Q_0 f_s(x^\sigma)\}_{(+)}, \qquad s \in S_0; \tag{6.4.16}$$

$$p_s^\sigma = \{\tilde{\tilde{p}}_s^\sigma + \alpha Q f_s(x^\sigma)\}_{(+)}, \qquad s \in S_i, \ i = 1, \ldots, I. \tag{6.4.17}$$

This algorithm is a parallel one-level analog of the general two-level algorithm described in Section 6.3. As in that algorithm, by selecting different values for the parameters Q_0', Q_0, Q, α, and by choosing different partitions of the set S, we can generate different variants of the algorithm.

Several observations are worth noting about this algorithm. Each subsystem computes not only its own preferred plan x^σ but also the plans $\Phi^{j,\sigma} z^{j,\sigma}$ that it recommends to the other subsystems. Similarly, each subsystem recommends to the other subsystems its shadow prices $\tilde{p}_s^{j,\sigma}$, $s \in S_i$, $i = 1, \ldots, I$, for the global resources. Then each subsystem determines the values (or price, that is, the value of the corresponding dual variable) of its local constraints with the values of the local constraints of the other subsystems fixed at the values determined at the previous iteration. It is this that makes it necessary to average over subsystems in (6.4.12)-(6.4.15). On the other hand, there is no need to average over the prices on the global constraints $s \in S_0$, as each subsystem uses these values equally in their objective function.

Equations (6.4.16), (6.4.17) arise from the fact that if a constraint is not satisfied at iteration σ, its price should be increased; if, however, the constraint is satisfied as a strict inequality, the value of the approximation should be decreased.

From the description of the aggregating functions we can see that

$$\varphi_k^{j,\sigma}(z_1^\sigma, \ldots, z_R^\sigma) = \sum_{r=1}^{R_j} \alpha_{rk}^{j,\sigma} z_r^j \qquad \text{and} \qquad \sum_{r=1}^{R} \alpha_{rk}^{j,\sigma} = \xi_k^{\sigma-1}.$$

Let the coefficients $\alpha_{rk}^{j,\sigma}$ be subject to the following conditions:

$$\sum_{r=1}^{R} \alpha_{rk}^{j,\sigma} = \begin{cases} \xi_k & \text{for } k \in K_i,\ i < j; \\ \xi_k^{\sigma-1} & \text{for } k \in K_i,\ i > j. \end{cases}$$

Proceeding in a similar manner, we can construct a sequential one-level algorithm. In solving its own problem, subsystem j determines x^σ and p_t^σ, $t \in L_j$. In constructing the aggregate objective function, it uses the previously computed values for the local constraints:

$$c_j^\sigma(x_j, z^j) = c(x_j, \Phi^{j,\sigma} z^j) - \sum_{i<j} \sum_{t \in L_i} p_t^\sigma h_t(\Phi_i^{j,\sigma} z)$$
$$- \sum_{i>j} \sum_{t \in L_i} p_t^{\sigma-1} h_t(\Phi_i^{j,\sigma} z). \tag{6.4.18}$$

All other operations for subsystem j are as in the parallel algorithm.

By the time we have solved all J of the local problems we will have determined every x_j^σ, p_t^σ, $t \in L_j$, $j = 1, \ldots, J$. The prices of the global constraints $s \in S_0$ are determined from (6.4.16), and for $s \in S_i$, $i = 1, \ldots, I$, from

$$p_s^\sigma = \{\lambda_i^{j,\sigma} p_s^{\sigma-1} + \alpha Q f_s(x^\sigma)\}, \quad s \in S_i,\ i = 1, \ldots, I. \tag{6.4.19}$$

We should point out that the order in which the subsystems solve their problems can vary from iteration to iteration.

6.5 GENERAL SCHEME OF TWO-LEVEL ITERATIVE AGGREGATION ALGORITHMS FOR CONSTRAINED OPTIMIZATION PROBLEMS

In this section we present a detailed analysis of a general approach to the construction of iterative aggregation algorithms that we first outlined in the Introduction. Just as for product balances, a successive realization of the major concepts of iterative aggregation enables us to generate a theory rather than an amalgam of separate and seemingly unrelated algorithms. Within the framework of this theory we will be able to accommodate both old and new algorithms as well as bring out the relative advantages and disadvantages of each algorithm.

The fundamental idea in two-level iterative aggregation algorithms is that the original problem is solved by decomposing it into subproblems that are solved at different levels of the control hierarchy. Each iteration σ consists of two computational stages. The first stage, which we call the "aggregation transformation" following Martines-Soler and Chernyak (1974), involves the following operations. Each subsystem computes its aggregate indices on the basis of the approximate solutions obtained at iteration $\sigma - 1$. These aggregate indices are then passed to the center. The center forms an aggregate problem using the indices, solves it, and passes the solution of this problem back to the subsystems. The subsystems then use the aggregate solution to adjust the approximation from iteration $\sigma - 1$.

Although aggregation transformations give some adjustment to the approximate solution from the preceding iteration, this adjustment turns out to be insufficient for two-level iterative aggregation algorithms to converge to the true solution. For these algorithms the second stage must generate a convergent sequence of approximations, and this requires the use of detailed information about the subsystems. The problems solved at this second stage are the local problems of the subsystems, and as a rule represent one iteration or part of one iteration of an iterative solution process, which may be a block process.

The solution scheme for a two-level process is illustrated in Figure 6.1. Formally, the scheme can be described as follows. Let the aggregation transformation be denoted by the operator $F(x)$ and let $G(x)$ be the operator equivalent to the second stage; that is, it corresponds to the solution of the local problems. Then

$$x^{\sigma} = G(F(x^{\sigma-1})), \tag{6.5.1}$$

where $x^{\sigma-1}$, x^{σ} are the vectors of variables obtained at iteration $\sigma - 1$ and σ, respectively. Letting $\tilde{x}^{\sigma}$ denote $F(x^{\sigma-1})$, then

$$x^{\sigma} = G(\tilde{x}^{\sigma}). \tag{6.5.2}$$

What defines a specific iterative aggregation algorithm is the choice of the adjustment:

$$\tilde{x}^{\sigma} = F(x^{\sigma-1}). \tag{6.5.3}$$

Therefore, two-level iterative aggregation processes can usually be said to consist of incorporating the adjustment (6.5.3) into some algorithm for solving the original problem:

$$x^{\sigma} = G(x^{\sigma-1}). \tag{6.5.4}$$

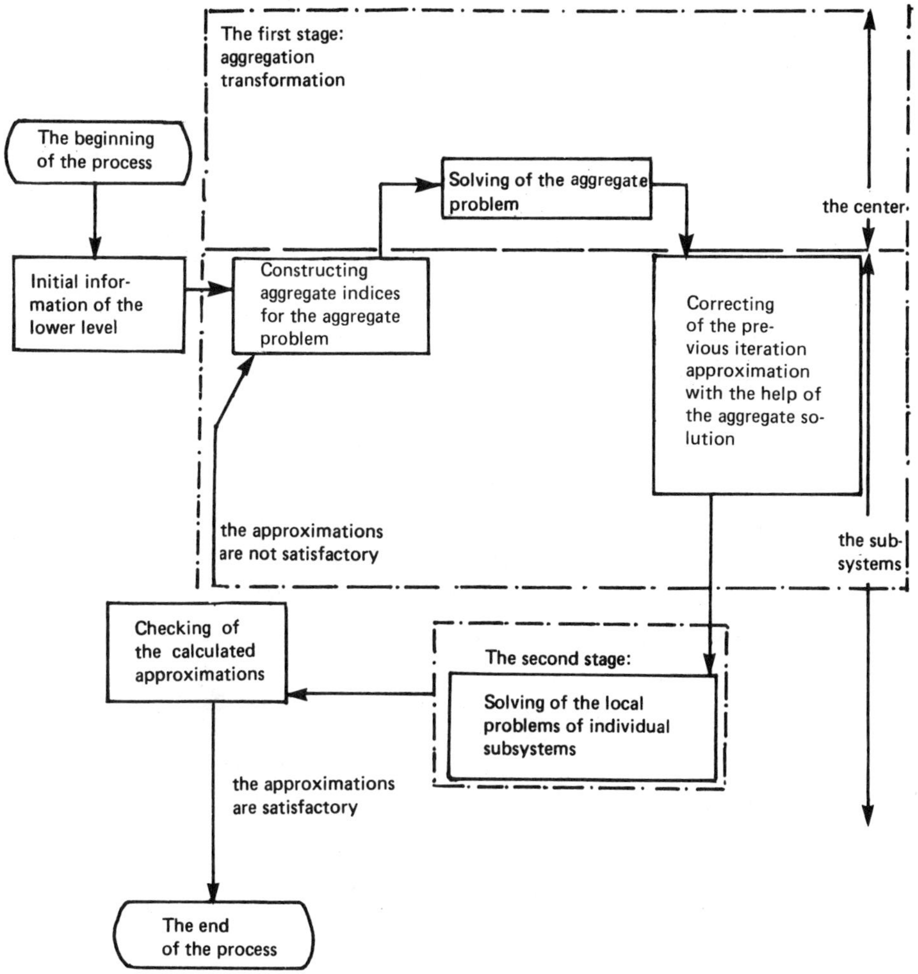

Figure 6.1. Scheme of a two-level iterative aggregation process.

In some algorithms it is possible to make do with just the first adjustment when the aggregate problem has a more complex structure which takes into account more information about the subsystems.

Let us now analyze the two-level iterative aggregation algorithms for solving extremum problems using Lagrangian functions proposed in Section 6.4. For ease of exposition we will only consider problems that have inequality constraints given by

$$f_s(\xi_1,\ldots,\xi_K) \leq 0, \qquad s \in S; \tag{6.5.5}$$

$$x \in Q; \tag{6.5.6}$$

$$\max\ c(\xi_1,\ldots,\xi_K). \tag{6.5.7}$$

As in Section 6.3, assume that the set $\{1,\ldots,K\}$ is partitioned into J subsets K_j, $j = 1, \ldots, J$, and that the set S is partitioned into I disjoint subsets S_i, $i = 1, \ldots, I$. Let x_j denote the vector $(\xi_k | k \in K_j)$ and y_i the vector $(p_s | s \in S_i)$. We restrict p_s to be greater than or equal to zero since the constraints (6.5.5) are all inequalities. Similarly, $x_j^\sigma = (\xi_k^\sigma | k \in K_j)$ and $y_i^\sigma = (p_s^\sigma | s \in S_i)$.

In the iterative aggregation algorithms discussed previously for solving (6.5.5)-(6.5.7) variable aggregation consisted of fixing the proportions between ξ_k and ξ_i where both k and i belonged to K_j. Thus for any given vector ξ we can pass from the functions $c(\xi_1,\ldots,\xi_k)$ and $f(\xi_1,\ldots,\xi_k)$ to the functions $\hat{c}_\xi(z_1,\ldots,z_J)$ and $f_\xi(z_1,\ldots,z_J)$ by the substitution $z_j\xi_k$, $k \in K_j$, $j = 1, \ldots, J$. This effects a transformation from the variables ξ_k, $k = 1, \ldots, K$ to the variables z_j, $j = 1, \ldots, J$. Then the constraint $\xi \in Q$ is transformed into the constraint $z \in T$, where $T = \{z = (z_1,\ldots,z_J) | (z_1x_1, \ldots, z_Jx_J) \in Q\}$. Note that if $Q = \mathbb{R}_+^K$, then $T = \mathbb{R}_+^J$.

Next we consider the problem of how to aggregate the constraints (6.5.5). In some algorithms the constraints are not aggregated so that the aggregate problem has constraints of the form

$$\hat{f}_{x,s}(z_1,\ldots,z_J) = 0, \qquad s \in S. \tag{6.5.8}$$

In this case the center has to adjust the vectors x_j so as to obtain the maximum value of the objective function possible from such an adjustment while maintaining the feasibility of the solution vector. The aggregate problem that achieves this adjustment is

$$\hat{f}_s^\sigma(z) \leq 0, \qquad s \in S; \tag{6.5.9}$$

$$z \in T; \tag{6.5.10}$$

$$\max\ \hat{c}^{\sigma-1}(z). \tag{6.5.11}$$

Here $\hat{f}_s^{\sigma-1}(z) = \hat{f}_{x^{\sigma-1},s}(z)$, $\hat{c}^{\sigma-1}(z) = \hat{c}_{x^{\sigma-1}}(z)$. If $z^\sigma = (z_1^\sigma,\ldots,z_J^\sigma)$

is the solution of (6.5.9)-(6.5.11), it follows from the definition of the aggregation operator for the functions f_s and the set T that the vector $\xi^\sigma = (z_1^\sigma x_1^{\sigma-1}, \ldots, z_J^\sigma x_J^{\sigma-1})$ satisfies the constraints (6.5.5)-(6.5.7).

However, some choices for constructing the aggregate problem render (6.5.9)-(6.5.11) insolvable, and this is especially likely if $x^{\sigma-1}$ happens to be close to the vector x^*, the optimal solution vector of (6.5.5)-(6.5.7), which itself lies on the boundary of the set given by the constraints (6.5.5), (6.5.6).

Next we consider algorithms in which the aggregate problem contains fewer constraints than the original problem. All constraints will be considered to have the economic interpretation of resource constraints. To aggregate the constraints (6.5.5) we can construct a Lagrangian function which includes a term that corresponds to these constraints. The economic interpretation of this is that the master program (the center's problem) consists in determining the multipliers z_i which guarantee the maximum profit under given fixed prices on the nonbinding resources. In this case the aggregate problem has the form

$$z \in T; \tag{6.5.12}$$

$$\max\left\{\hat{c}^{\sigma-1}(z) - \sum_{s \in S} p_s^{\sigma-1}\hat{f}_s^{\sigma-1}(z)\right\}. \tag{6.5.13}$$

As distinct from (6.5.9)-(6.5.11), this problem is always solvable, provided that $x^{\sigma-1} \in Q$. However, even this problem has one considerable shortcoming: It can have too many solutions. Namely, let $x^{\sigma-1} = x^*$, $p^{\sigma-1} = p^*$ be the primal and dual solutions of (6.5.5)-(6.5.7). It is clear that in this case $z = (1,\ldots,1)$ is the solution to (6.5.12), (6.5.13). However, (6.5.12), (6.5.13) can have other solutions as well such that the vector $(z_1^\sigma x_1^{\sigma-1}, \ldots, z_J^\sigma x_J^{\sigma-1})$ no longer is a solution of the original problem (6.5.5)-(6.5.7).

Also, this method of forming the aggregate problem is not quite in keeping with the main idea of iterative aggregation, which is that when solving the aggregate problem we allow the variables of original problem to vary in accordance with certain rules (e.g., in the algorithms presented so far we have allowed them to proportionally increase or decrease). In the problem (6.5.12), (6.5.13), however, only the primal variables of the original problem possess this property, while the dual variables remain fixed. Therefore, it is natural to attempt to construct the upper-level problem to fix the proportions between the dual variables rather than their absolute values. Since this requires determining a magnitude of variation for the dual variables, the aggregate problem should contain constraints of the form

$$\sum_{s \in S_i} p_s^{\sigma-1} \hat{f}_s^{\sigma-1}(z) \leq 0, \qquad i = 1, \ldots, I. \tag{6.5.14}$$

Thus if it is desired to fix the values of $p_s^{\sigma-1}$, $s \in S_0$ at iteration σ and to allow the values of the dual variables in constraint subsets S_i, $i = 1, \ldots, I$ to change proportionally, we can represent the upper-level problem in the following form:

$$\sum_{s \in S_i} p_s^{\sigma-1} \hat{f}_s^{\sigma-1}(z) \leq 0, \qquad i = 1, \ldots, I; \tag{6.5.15}$$

$$z \in T; \tag{6.5.16}$$

$$\max \left\{ \hat{c}^{\sigma-1}(z) - \sum_{s \in S_0} p_s^{\sigma-1} \hat{f}_s^{\sigma-1}(z) \right\}. \tag{6.5.17}$$

Since the solution of (6.5.15)-(6.5.17) is represented by the saddle point of the Lagrangian function

$$L(z,\lambda) = \hat{c}^{\sigma-1}(z) - \sum_{s \in S_0} p_s^{\sigma-1} \hat{f}^{\sigma-1}(z) - \sum_{i=1}^{I} \lambda_i \sum_{s \in S_i} p_s^{\sigma-1} \hat{f}^{\sigma-1}(z), \qquad \lambda_i \geq 0,\ z \in T, \tag{6.5.18}$$

over the set $z \in T$. We can see that elements of the dual solution $\lambda^\sigma = (\lambda_1^\sigma, \ldots, \lambda_I^\sigma)$ of (6.5.15)-(6.5.17) indicate by what factor we have to change the values of $p_s^{\sigma-1}$ in order to obtain the optimum of the upper-level problem for given proportions within the groups of the primal and dual variables. Martines-Soler (1977) used (6.5.15)-(6.5.17) with S_0 the empty set to solve nonlinear problems.

The problem (6.5.15)-(6.5.17) also has several considerable shortcomings. Just as with the problem (6.5.9)-(6.5.11) it may be insolvable for some $x^{\sigma-1}$, although the likelihood of this is much smaller than in the previous algorithm. Indeed, for some values of $x^{\sigma-1}$ there may not exist a vector z that satisfies all the constraints (6.5.15), (6.5.16). We can overcome this difficulty by computing the vector $\hat{x}^{\sigma-1} = (z_1 x_1^{\sigma-1}, \ldots, z_J x_J^{\sigma-1})$ that is closest to the set defined by the constraints (6.5.5), (6.5.6), and then determining the projection $\hat{\hat{x}}^{\sigma-1}$ of this point onto the set. If we now construct the aggregate problem on the basis of either $x^{\sigma-1}$ or $\hat{\hat{x}}^{\sigma-1}$, it will prove to be solvable since any problem (6.5.15)-(6.5.17) constructed on the basis of $x^{\sigma-1}$ that satisfies (6.5.5), (6.5.6) is also solvable. However, this projection will often mean solving a problem that is comparable in complexity to the original problem. In the general case these projections can adversely affect convergence. The use of the projections does ensure that the approximate solution obtained at each iteration is feasible, but this advantage hardly seems to outweigh the disadvantages mentioned previously.

We now turn to a more fundamental shortcoming with the problems (6.5.12), (6.5.13), and (6.5.15)-(6.5.17). For arbitrary $x^{\sigma-1}$, $p^{\sigma-1}$ the problems (6.5.12), (6.5.13), and (6.5.15)-(6.5.17) may happen to have nonunique solutions, and since we do not know which of these values of z^{σ} to use, the process is not necessarily convergent in the general case. This shortcoming is a much more serious problem since it relates to the properties of the classical Lagrangian function. It is quite natural that an attempt to overcome this problem will lead us to modified Lagrangian functions (see, e.g., Hestenes, 1969; Powell, 1969; Golstein, 1972; Polyak and Tretyakov, 1973; Fletcher, 1973; Tretyakov, 1973; Rockafellar, 1973; Golstein and Tretyakov, 1974; Razumikhin, 1975) which can be used in two different ways to construct the aggregate problem: to aggregate the problem first and then construct a modified Lagrangian function for the aggregate problem or to construct a modified Lagrangian function for the original problem first and then apply the aggregation operator.

It is clear that the first of these methods, forming the Lagrangian function of the aggregate problem, can only be applied to the problem (6.5.15)-(6.5.17). If we use the standard Lagrangian function modified by quadratic penalties, we can obtain either one of two different problems. The first of these problems has the form

$$z \in T; \tag{6.5.19}$$

$$\max \Bigg\{ \hat{c}^{\sigma-1}(z) - \sum_{s \in S_0} p_s^{\sigma-1} \hat{f}_s^{\sigma-1}(z) - \sum_{i=1}^{I} \frac{1}{2q_i} \Bigg[\Bigg(\Bigg(\lambda_i + q_i \sum_{s \in S_i} p_s^{\sigma-1} f_s^{\sigma-1}(z) \Bigg)_{(+)} \Bigg)^2 - \lambda_i^2 \Bigg] \Bigg\}. \tag{6.5.20}$$

Here q_i is the penalty coefficient for the ith constraint in (6.5.15) and λ_i are the dual variables for these constraints.

As we can see, the values of the dual variables in this problem are fixed, and as has been noted previously, this is not in accordance with the main idea of iterative aggregation.

In the general case, however, any solution of (6.5.15)-(6.5.17) will also be a solution to (6.5.19), (6.5.20). Therefore, using (6.5.19), (6.5.20) instead of (6.5.15)-(6.5.17) does not entail any advantages as far as eliminating extra solutions. On the other hand, we know how to modify the aggregate problems so as to avoid generating problems with nonunique solutions, at least in the case where the problems are aggregated using x*, p*—the optimal primal and dual solutions, respectively. To do this it is sufficient to add to the aggregate functional the term

$$- \sum_{j=1}^{J} \frac{q_i}{2} (z_j - 1)^2 \tag{6.5.21}$$

With this modification the aggregate problem corresponding to (6.5.12), (6.5.13), that is, the aggregate problem in which the dual variables are held at fixed values, has the form

$$z \in T; \tag{6.5.22}$$

$$\max\left\{\hat{c}^{\sigma-1}(z) - \sum_{s\in S} p_s^{\sigma-1}\hat{f}_s^{\sigma-1}(z) - \sum_{j=1}^{J} \frac{q_j}{2}(z_j - 1)^2\right\}. \tag{6.5.23}$$

The modification specified by (6.5.21) combined with a reasonable choice of the operator G will guarantee the optimality of the stationary point of the process. However, for nonoptimal $x^{\sigma-1}$, $p^{\sigma-1}$, this modification slows the process down, as mentioned in Section 6.3. This modification can be given the economic interpretation that the center should not substantially change the variants of the plan recommended by the subsystems.

The modified problem that corresponds to (6.5.15)-(6.5.17), that is, the problem that allows the dual variables to be proportionally adjusted, assumes the form

$$\sum_{s\in S_i} p_s^{\sigma-1}\hat{f}_s^{\sigma-1}(z) \leq 0, \qquad i = 1, \ldots, I; \tag{6.5.24}$$

$$z \in T; \tag{6.5.25}$$

$$\max\left\{\hat{c}^{\sigma-1}(z) - \sum_{s\in S_0} p_s^{\sigma-1}\hat{f}^{\sigma-1}(z) - \sum_{j=1}^{J} \frac{q_i}{2}(z_j - 1)^2\right\}. \tag{6.5.26}$$

Just as with the previous problem, this modification eliminates multiple solutions, but it does not eliminate the possibility that the problem is not solvable.

We now turn to the second method for using Lagrangian functions in which we construct the Lagrangian function for the original problem and then aggregate this function. Actually, this second method consists in constructing the aggregate problems (6.5.12), (6.5.13), and (6.5.15)-(6.5.17) not for (6.5.5)-(6.5.7) but for the following equivalent problem:

$$f_s(x) \leq 0, \qquad s \in S; \tag{6.5.27}$$

$$x \in Q; \tag{6.5.28}$$

$$\max_x\left\{c(x) - \sum_{s\in S} \frac{q_s}{2} f_s^2(x)_{(+)}\right\}. \tag{6.5.29}$$

Following the suggestions of Golstein and Tretyakov (1974), we can construct other modifications to the aggregate problem. However, (6.5.29) is the most widely used modification and it is also sufficient to illustrate the essential ideas we want to present.

If we construct the aggregate problem (6.5.12), (6.5.13) for (6.5.27)-(6.5.29) when J = K, we will obtain

$$x \in Q; \tag{6.5.30}$$

$$\max_x \left\{ c(x) - \sum_{s \in S} \frac{1}{2q_s} [((p_s^{\sigma-1} + q_s f(x))_{(+)})^2 - (p_s^{\sigma-1})^2] \right\}. \tag{6.5.31}$$

If, however, J < K, we obtain the problem in which the functional (6.5.31) has to be maximized only on some subset of $x \in Q$ and not on the whole set. In this case the problem assumes the form

$$z \in T; \tag{6.5.32}$$

$$\max \left\{ \hat{c}^{\sigma-1}(z) - \sum_{s \in S} \frac{1}{2q_s} [(p_s^{\sigma-1} + q_s \hat{f}_s(z))_{(+)}]^2 \right\}. \tag{6.5.33}$$

Using (6.5.27)-(6.5.29), we can construct a problem analogous to (6.5.15)-(6.5.17) in which the values of both the primal and dual variables will be adjusted. The aggregate problem in this case will have the form

$$\sum_{s \in S_i} p_s^{\sigma-1} \hat{f}_s^{\sigma-1}(z) \leqq 0, \qquad i = 1, \ldots, I; \tag{6.5.34}$$

$$z \in T; \tag{6.5.35}$$

$$\max_z \left\{ \hat{c}^{\sigma-1}(z) - \sum_{s \in S_0} \frac{1}{2q_s} [(p_s^{\sigma-1} + q_s \hat{f}_s^{\sigma-1}(z))_{(+)}]^2 \right\}. \tag{6.5.36}$$

The problem (6.5.34)-(6.5.36) most fully reflects the ideas of iterative aggregation, but it is subject to the same reservations concerning solvability as noted for the algorithm (6.5.15)-(6.5.20).

We now give an economic interpretation to the adjustment using aggregate problems. Recall that the saddle point of the Lagrangian function (of either the classical or modified Lagrangian) can be viewed as the equilibrium point between output plans and resource prices. In this sense the solution of the aggregate problem amounts to a search for the equilibrium between plans and resource prices provided that both the plans and the resource prices lie in a specified cone.

This method of interpreting aggregation-based adjustment enables us to construct an aggregate problem in which a term similar to (6.5.21) is used to damp not only the primal variables but the dual prices as well. For the general nonlinear case the aggregate problem then assumes the form

$$z \in T; \tag{6.5.37}$$

$$\min_{\lambda} \max_{z} \left\{ \hat{c}^{\sigma-1}(z) - \sum_{i=0}^{I} \lambda_i \sum_{s \in S_i} p_s^{\sigma-1} \hat{f}_s^{\sigma-1}(z) - \sum_{j=1}^{J} \frac{q_i}{2} (z_j - 1)^2 + \sum_{i=0}^{I} \frac{\tilde{q}_i}{2} (\lambda_i - 1)^2 \right\}. \tag{6.5.38}$$

The problem is thus a search for the saddle point of the Lagrangian function. We believe that the term $\Sigma_{i=0}^{I} (\tilde{q}_i/2)(\lambda_i - 1)^2$ should make the process more stable with respect to the dual variables. Note that this problem has a solution for all $x^{\sigma-1}$.

The idea of having a minimax problem as the aggregate problem was first proposed by Telle (1977, Chap. 3), although he uses a method of aggregation which possesses a number of disadvantages compared to the methods used in this book. Telle's method is only applicable to the case where the functions (6.5.5), (6.5.7) are separable with respect to the groups of variables $x_j = \{\xi_k | k \in K_j\}$:

$$c(x) = \Sigma_j c_j(x_j);$$

$$f_s(x) = \Sigma_j f_{sj}(x_j), \qquad s \in S.$$

and the aggregation functions he suggests are always linear of the form

$$c^{\sigma-1}(z) = \Sigma_j z_j c_j(x_j^{\sigma-1}),$$

$$f_s^{\sigma-1}(z) = \Sigma_j z_j f_{sj}(x_j^{\sigma-1}), \qquad s \in S.$$

Thus it is not really the variables that are being aggregated but rather the values of the functions. If all the functions c_j and f_{sj} are homogeneous of the same degree, this aggregation procedure will coincide with the one used in this book; in other cases, however, it may lead to considerable loss of accuracy. Although the problem (6.5.37), (6.5.38) theoretically is always solvable, finding the actual solutions may present considerable difficulties since methods for finding saddle points of functions are not at present as well developed as methods for solving constrained optimization problems.

Thus suppose that a certain aggregate problem has been selected. Then as a result of its solution at iteration σ we obtain z_j^σ, $j = 1, \ldots, J$, and possibly λ_i^σ, $i = 1, \ldots, I$. Following the proposed methodology, we would then adjust the present approximation by

$$\xi_k = z_j^\sigma \xi_k^{\sigma-1}, \qquad k \in K_j, \qquad j = 1, \ldots, J, \tag{6.5.39}$$

and if λ_i^σ has also been calculated, then

$$\tilde{p}_s^\sigma = \lambda_i^\sigma p_s^{\sigma-1}, \qquad s \in S_i, \qquad i = 1, \ldots, I. \tag{6.5.40}$$

If some z_j or λ_i have not been computed, then of course the corresponding adjustment is not made. The formulas (6.5.39), (6.5.40) end the sequence of transformations that characterize the operator F from (6.5.3). As to the beginning of this sequence, we have already shown that it may involve several different methods for constructing the aggregate problem.

We pointed out previously that the adjustment (6.5.39), (6.5.40) cannot by itself ensure that the algorithm will converge. Therefore, to determine x^σ, p^σ we have to perform one or several iterations of some other algorithm, that is, to realize the operator G from (6.5.2). Thus Martines-Soler (1973) in his iterative aggregation algorithm uses as G one iteration of the generalized gradient method for determining the saddle point (see Golstein, 1972):

$$x^\sigma = G(\tilde{x}^{\sigma-1}) = x^\sigma + \alpha^\sigma \nabla_x L(\tilde{x}^\sigma, \tilde{p}^\sigma), \tag{6.5.41}$$

$$p^\sigma = G(\tilde{p}^\sigma) = \{p^{\sigma-1} - \alpha^\sigma \nabla_p L(\tilde{x}^\sigma, \tilde{p}^\sigma)\}, \qquad (6.5.42)$$

where the numerical sequence α^σ satisfies the conditions

$$\alpha^\sigma > 0, \qquad \lim_{\sigma\to\infty} \alpha^\sigma = 0, \qquad \sum_\sigma \alpha^\sigma = \infty, \qquad (6.5.43)$$

and $\nabla_x L(\tilde{x}^\sigma, \tilde{p}^\sigma)$ and $\nabla_p L(\tilde{x}^\sigma, \tilde{p}^\sigma)$ are the vectors of partial derivatives evaluated at the point $\tilde{x}^\sigma$, $\tilde{p}^\sigma$ of the modified Lagrangian function (6.5.38).

In the algorithm of Dudkin and Vakhutinsky described in Section 6.3, the aggregation operator F includes the problem (6.5.34)-(6.5.36), which has an objective function augmented by a term of the form (6.5.21) and the functions f_s^σ that are included in the objective function are replaced by their linear approximations. After solving the aggregate problem and adjusting the solution by (6.5.42), (6.5.43) we perform one iteration of the Bakhmetyev et al. (1973) decomposition algorithm in order to obtain x^σ and p^σ (i.e., to realize the operator G). The necessary generalization of this algorithm to the nonseparable case in question differs from the original algorithm in that the variables of the other subsystems in the jth local problem are fixed at a level that corresponds to the values from the preceding iteration.

Thus we have presented a general overview of the presently existing iterative aggregation algorithms. If we follow the methodology established in the previous chapters, it would seem natural to continue the development of these algorithms by aggregating increments to the variables rather than the variables themselves. This would enable us to make the algorithms more flexible and utilize the solution of the aggregate problem with greater effectiveness.

As in Chapter 5, we shall deal exclusively with linear aggregation; that is, the aggregation of F(x) at the point $x^{\sigma-1}$ with respect to the VDIs $\delta^{j,\sigma-1}$, $j = 1, \ldots, J$, will be understood to entail a transition to F(z) by replacing x in F(x) by $x + \Sigma_{j=1}^J z_j \delta^{j,\sigma-1}$. In this type of aggregation we can interpret the quantities z_j as the magnitude of increase of variables in the proportions $\delta_k^{j,\sigma-1}$. When the primal variables are aggregated in this manner but the dual variables are aggregated as before, the quadratic term (6.5.21) in the aggregate problem expressed in terms of (6.5.23), (6.5.26), (6.5.38) has to be replaced by

$$-\sum_j \frac{q_j}{2} z_j^2, \tag{6.5.44}$$

and the adjustment (6.5.39) realized by the formula

$$\tilde{\xi}_k^{\sigma} = \xi_k^{\sigma-1} + \Sigma_j\ z_j^{\sigma}\delta_k^{j,\sigma-1}. \tag{6.5.45}$$

As in Chapter 5, we obtain the original algorithms when $\delta_k^{j,\sigma-1} = \xi_k^{\sigma-1}$ for $k \in K_j$, $\delta_k^{j,\sigma-1} = 0$ for $k \notin K_j$. It would seem natural, for example, to let $\delta^{j,\sigma-1}$ be the vector of partial derivatives of the Lagrangian function (or a modified Lagrangian function) of the original problem at the point $(x^{\sigma-1},p^{\sigma-1})$. Generally speaking, however, a good method for choosing the vector $\delta^{j,\sigma-1}$ at iteration σ constitutes a separate problem in its own right, which will not be discussed here.

Note that using block aggregation, as defined in Chapter 5, where the set Q from (6.5.6) is a parallelepiped, makes it rather simple to describe the set T in (6.5.10). In this case T is also a parallelepiped. This can give block aggregation an advantage over the more general form of aggregation in actual computations since in the more general form of aggregation the set T may prove to be so complex that the computational cost of solving the resulting aggregate problem would nullify the gains obtained from a better choice of the VDIs $\delta^{j,\sigma-1}$.

Thus conceptually, general linear variable aggregation in constrained optimization problems is not at all different from general linear variable aggregation in unconstrained optimization problems. This is not the case for dual variable aggregation, as unconstrained optimization problems do not have dual variables. Recall that in the previous aggregation methods (without aggregation of increments), the main idea of the upper-level problem consists of determining the saddle point of the Lagrangian function (or a modified Lagrangian function) when both primal and dual variables run through some homogeneous cones. "General linear aggregation" means that instead of homogeneous cones we take arbitrary linear manifolds. However, it we take nonhomogeneous manifolds in the space of dual variables, we have to restrict ourselves to problems that have only equality constraints.

These considerations lead us to the following unified method of aggregating dual variables. Let the original problem contain the constraints

$$f_s(x) = 0, \qquad s \in S, \tag{6.5.46}$$

and let the saddle point be determined such that $p_s = p_s^{\sigma-1} + \Sigma_i \lambda_i \theta_s^{i,\sigma-1}$. Then to aggregate the dual variables we have to add the term

$$-\sum_{s\in S} p_s^{\sigma-1} f_s(x), \tag{6.5.47}$$

to the functional and the aggregate constraint corresponding to the ith VDI $\theta^{i,\sigma-1}$ will be

$$\sum_{s\in S} \theta_s^{i,\sigma-1} f_s(x) = 0. \tag{6.5.48}$$

Thus if (6.5.15)-(6.5.17) is our aggregate problem, general linear dual variable aggregation will change it to

$$\sum_{s\in S} \theta_s^{i,\sigma-1} \hat{f}_s^{\sigma-1}(z) = 0, \qquad i = 1, \ldots, I; \tag{6.5.49}$$

$$z \in T; \tag{6.5.50}$$

$$\max_z \left\{ \hat{c}^{\sigma-1}(z) - \sum_{s\in S} p_s^{\sigma-1} \hat{f}_s^{\sigma-1}(z) \right\}. \tag{6.5.51}$$

The problems (6.5.22), (6.5.23), (6.5.24)-(6.5.26), and (6.5.34)-(6.5.36) are transformed in the same manner as (6.5.15)-(6.5.17).

Because of the special importance of the aggregate problem

(6.5.34)-(6.5.36), we describe the problem that uses aggregation of increments that corresponds to it:

$$\sum_{s \in S_i} \theta_s^{i,\sigma-1} \hat{f}_s^{\sigma-1}(z) = 0, \qquad i = 1, \ldots, I; \tag{6.5.52}$$

$$z \in T; \tag{6.5.53}$$

$$\max_z \left\{ \hat{c}^{\sigma-1}(z) - \sum_{s \in S} p_s^{\sigma-1} \hat{f}_s^{\sigma-1}(x) - \sum_{s \in S_0} \frac{q_s}{2} [\hat{f}_s^{\sigma-1}(x)]^2 \right\}. \tag{6.5.54}$$

In conclusion, we note that for each of the aggregate algorithms presented in this section there are cases when this algorithm is the most effective. For example, if $c(x)$ is strictly convex and $f_s(x)$ are linear, classical Lagrangian functions may be enough and we can use the simplest aggregate problem (6.5.49)-(6.5.51). On the other hand, the most complex problem (6.5.52)-(6.5.54) enables us to construct effective algorithms for the widest variety of problems. The aggregate problem generated by (6.5.24)-(6.5.26) is simpler than (6.5.52)-(6.5.54), but for the algorithm based on this problem it is clearly not enough to calculate the new approximations of the primal and dual variables just using the solution of the aggregate problem. It seems to us that using aggregation of increments, despite the increased difficulty in constructing the problem, makes it possible to generate considerably more effective iterative aggregation algorithms.

Also note that Mendelssohn (1982) proposed a two-level iterative aggregation algorithm for the linear programming problem of a particular type. This specific linear programming problem corresponds to the Markov decision processes. The two-level iterative aggregation algorithm for this problem uses simple aggregation of both primal variables and constraints and constructing master program (6.5.15)-(6.5.17) with $S_0 = \emptyset$. New approximations are determined from solutions of the local linear programming problems. Mendelssohn proves convergence of this algorithm.

6.6 GENERAL SCHEME OF ONE-LEVEL ITERATIVE AGGREGATION ALGORITHMS FOR CONSTRAINED OPTIMIZATION PROBLEMS

The theory set forth in Section 6.5 suggests the following way of applying the theory to one-level algorithms in general; namely, some of the variables in any of the aggregate problems presented in the preceding section are not to be aggregated. That is, some subsets K_j will contain only one element and the union of these subsets for a given subsystem will include all its variables. Thus, in one-level algorithms each subsystem solves an aggregate problem that has been specifically constructed for it. Therefore, everything in the preceding section concerning the aggregate problems can be applied *in toto* to the local problems of one-level algorithms. An informal description of the difference between two- and one-level algorithms is as follows. In one-level algorithms the subsystem determines not only the vector x_j, which contains its own variables, but also the aggregate variables z_i^j that correspond to the aggregate variables of the other subsystems.

Thus, if some subsystems, say the first n, solve their problems in parallel, each vector x_j will have n values. If for each subsystem j we choose as x_j^σ those vectors that the subsystems have determined for themselves, the aggregate vector x^σ can turn out to be much worse in many respects than the vector $x^{\sigma-1}$. Hence to obtain x_j^σ we have to average over the recommendations of all the subsystems with respect to the values of this vector. We should point out that all these remarks apply equally to the price vectors, which means we have to use the formulas (6.4.12)-(6.4.14).

We do not have to resort to averaging if the subsystems solve their problems sequentially, that is, when each subsystem uses the most recent solutions of the other subsystems to obtain its own solution. Note that one step of the one-level process can be used for calculating the new approximations from the solution of the upper-level problem in a two-level algorithm.

We now describe a general form of the algorithm in question which includes both one- and two-level iterative aggregation algorithms. Due to the fact that aggregate problems can be constructed not only for the variables themselves but also for their increments, we shall analyze only this most general case. Thus, suppose that for each σ we have an array of functions ($t = 1, \ldots, m$):

$$\varphi_k^{t,\sigma}(z_1,\ldots,z_{R_t}), \qquad k = 1, \ldots, K; \tag{6.6.1}$$

$$\Phi_s^{t,\sigma}(\lambda_1,\ldots,\lambda_{I_t}), \qquad s \in S. \tag{6.6.2}$$

Here R_t and I_t are some positive numbers. If all the functions are linear and $\varphi_k^{t,\sigma}(0) = \xi_k^{\sigma-1}$, $\Phi_k^{t,\sigma}(0) = p_s^{\sigma-1}$, then

$$\varphi_k^{t,\sigma}(z_1,\ldots,z_{R_t}) = \xi_k^{\sigma-1} + \sum_{r=1}^{R_t} z_r \delta_{tk}^{r,\sigma-1}; \tag{6.6.3}$$

$$\Phi_s^{t,\sigma}(\lambda_1,\ldots,\lambda_{I_t}) = p_s^{\sigma-1} + \sum_{i=1}^{I_t} \lambda_i \theta_{is}^{t,\sigma-1}. \tag{6.6.4}$$

Thus for each $t = 1, \ldots, m$ we have brought into our analysis R_t of the VDIs $\delta_{tk}^{r,\sigma-1}$ for the primal variables and I_t VDIs $\theta_{is}^{t,\sigma-1}$ for the dual variables. The limiting case described in Section 6.3 is obtained if we set $\delta_{tk}^{r,\sigma-1} = 0$ for $k \notin K_r$ and $\theta_{i,s}^{t,\sigma-1} = 0$ for $s \notin S_i$. Formally replacing ξ_k by $\varphi_k^{t,\sigma}(z_1,\ldots,z_{R_t})$ and p_s by $\Phi_s^{t,\sigma}(\lambda_1,\ldots,\lambda_I)$ in the corresponding Lagrangian functions, we obtain the aggregate problem. Thus the problem (6.5.52), (6.5.53) will assume the form

$$\sum_{s \in S_i} \theta_{is}^{t,\sigma-1} p_s^{\sigma-1} \hat{f}_{st}^{\sigma-1}(z_1,\ldots,z_{R_t}) \leq 0, \qquad i = 1, \ldots, I; \tag{6.6.5}$$

$$z \in T; \tag{6.6.6}$$

$$\max_z \left\{ \hat{c}_t^{-1}(z_1,\ldots,z_{R_t}) - \sum_{s \in S_0} p_s^{\sigma-1} \hat{f}_{st}^{\sigma-1}(z_1,\ldots,z_{R_t}) \right\}. \tag{6.6.7}$$

Here $\hat{f}_{st}^{\sigma-1}(z_1,\ldots,z_{R_t})$ is the function $f_s(\xi_1,\ldots,\xi_K)$ aggregated on the basis of (6.6.1) and $\hat{c}_t^{\sigma-1}(z_1,\ldots,z_{R_t})$ the function $c(\xi_1,\ldots,\xi_K)$ aggregated on the basis of (6.6.1).

Thus, by aggregating using (6.6.1), (6.6.2) we obtain m aggregate problems. Solving them in parallel we obtain the vectors $z^{t,\sigma}$, $\lambda^{t,\sigma}$, t = 1, ..., m, on the basis of which we carry out the adjustment

$$\xi_k^{t,\sigma} = \varphi_k^{t,\sigma}(z_1^{t,\sigma},\ldots,z_{R_t}^{t,\sigma}), \qquad k = 1, \ldots, K; \tag{6.6.8}$$

$$\tilde{p}_s^{t,\sigma} = \Phi_s^{t,\sigma}(\lambda_1^{t,\sigma},\ldots,\lambda_{I_t}^{t,\sigma}), \qquad s \in S,\ \sigma = 1, \ldots, m. \tag{6.6.9}$$

Thus for each t = 1, ..., m we calculate separate adjusted vectors $\xi^{t,\sigma}$, $p^{t,\sigma}$. Then we finally calculate the fully adjusted vectors:

$$\tilde{\xi}^{\sigma} = \sum_{t=1}^{m} \alpha_t^{\sigma}\tilde{\xi}^{t,\sigma}; \tag{6.6.10}$$

$$\tilde{p}^{\sigma} = \sum_{t=1}^{m} \alpha_t^{\sigma}\tilde{p}^{t,\sigma}; \tag{6.6.11}$$

$$\alpha_t^{\sigma} > \delta > 0, \qquad \sum_{t=1}^{m} \alpha_t^{\sigma} = 1. \tag{6.6.12}$$

If m = J and the aggregating functions are such that $\varphi_k^{j,\sigma}(z_1,\ldots,z_{R_t}) = \xi_k^{\sigma-1} + z_k^{\sigma-1}$ for some t, with $k \in K_j$ and assuming that all other $\varphi^{j,\sigma}(z_1,\ldots,z_{R_t})$ do not depend on z_1 (or as another possibility for all $k \in K_j$, either $\partial\varphi_k^{j,\sigma}/\partial z_1 = 0$ or $\partial\varphi_1^{j,\sigma}/\partial z_k = \xi_k^{\sigma-1}$); and analogously $\Phi_s^{j,\sigma}(\lambda_1,\ldots,\lambda_{I_t}) = p_s^{\sigma-1} + \lambda_i p_s^{\sigma-1}$ for some i, for $s \in S_i$ and all other $\Phi_s^{j,\sigma}(\lambda_1,\ldots,\lambda_{I_t})$ do not depend on λ_i, we obtain an aggregate problem in which we aggregate the variables of all subsystems except those of subsystem j. We assume that the set of constraints is also partitioned into subsets that only contain constraints that correspond to a single subsystem. In this case, by taking $\xi_k^{\sigma} = \tilde{\xi}_k^{\sigma}$, $p_s^{\sigma} = \tilde{p}_s^{\sigma}$ from (6.6.10), (6.6.11) we obtain a parallel one-level algorithm. Note that the choice of VDIs, that is, of the aggregating functions (6.6.1), (6.6.2), is carried out in accordance with the same principles as in the case of the two-level algorithms from Section 6.5.

We can also solve all the subsystems' problems sequentially and in constructing the problem of subsystem j take the approximations received already by other subsystems on the same iteration; that is, take $\tilde{\xi}_k^{t,\sigma}$ and $p_s^{t,\sigma}$ instead of $\xi_k^{\sigma-1}$, $p_s^{\sigma-1}$, then we get the sequential one-level algorithm.

Note that to obtain p_s^σ we can adjust $\tilde{p}_s^\sigma$ given by (6.6.11) in the following way:

$$p_s^\sigma = \tilde{p}_s^\sigma + \alpha f_s(\xi^\sigma). \tag{6.6.13}$$

It is always necessary to recompute p on the basis of (6.6.13) for those constraints whose corresponding dual variables were fixed during the iteration, that is, when $\partial\Phi_s^{t,\sigma}/\partial\lambda_i = 0$ for all i, t.

APPENDIX 6.1. ON THE CONVERGENCE OF ONE-SIDED ITERATIVE AGGREGATION ALGORITHMS (CONSTRAINTS AGGREGATION)

In this appendix we discuss convergence properties of iterative aggregation algorithms that use constraint aggregation to solve problems where all the constraints are equality constraints. As we will show later, our analysis of one-sided iterative aggregation algorithms with constraint aggregation for constrained optimization problems will lead us to solving the unconstrained minimization problem which is dual to the original problem. Therefore, to facilitate our references back to Chapter 1, where algorithms for solving unconstrained maximization problems were first examined, we shall assume that the original problem is one of constrained minimization. In this case the dual problem will be a maximization problem which corresponds to the problem of Chapter 1. Thus the problem to be investigated is

$$f_s(x) = 0, \qquad s \in S; \tag{A6.1.1}$$

$$x \in Q; \tag{A6.1.2}$$

$$\min\ c(x) \tag{A6.1.3}$$

In this appendix we will deal only with iterative aggregation algorithms in which the aggregate problem at iteration σ has the form

$$\min \left\{ c(x) + \sum_{s \in S} p_s^{\sigma-1} f_s(x) + \sum_{s \in S} \frac{q}{2} f_s^2(x) \right\}; \quad \text{(A6.1.4)}$$

$$\sum_{s \in S} \theta_{is} f_s(x) = 0, \qquad i = 1, \ldots, I; \quad \text{(A6.1.5)}$$

$$x \in Q. \quad \text{(A6.1.6)}$$

Here $p^{\sigma-1}$ is the approximation at iteration $\sigma - 1$ of the vector of dual variables that correspond to the constraints (A6.3.1) and $\theta_i = (\theta_{is} | s \in S)$, $i = 1, \ldots, I$, are the corresponding VDIs of these dual variables.

We will investigate the convergence of the resulting iterative aggregation algorithms by considering a functional dual to the problem (A6.1.1)-(A6.1.3). Recall that if $M(x,p)$ is the classical or some modified Lagrangian function of the problem (A6.1.1)-(A6.1.3), the dual function $\Psi(p)$ with respect to $M(x,p)$ is the function

$$\Psi(p) = \inf_{x \in Q} M(x,p). \quad \text{(A6.1.7)}$$

In what follows the modified Lagrangian function $M(x,p)$ of (A6.1.1)-(A6.1.3) will always represent the Lagrangian with quadratic penalty terms:

$$c(x) + \sum_{s \in S} p_s f_s(x) + \sum_{s \in S} \frac{q}{2} f_s^2(x), \quad \text{(A6.1.8)}$$

that is, the classical Lagrangian function $L(x,p)$ augmented by the penalty term

$$\sum_{s \in S} \frac{q}{2} f_s^2(x). \quad \text{(A6.1.9)}$$

Here the number q is the penalty coefficient. Let $\psi(p)$ denote the dual function [with respect to $M(x,p)$] of (A6.1.1)-(A6.1.3). If $c(x)$ is convex and $f_s(x)$, $s \in S$, are linear, the function $\psi(p)$ has a number of useful properties which are described in the next proposition.

PROPOSITION A6.1.1 Let $c(x)$ be a continuous concave function and $f_s(x)$, $s \in S$, linear functions. Then

1. $\psi(p)$ is convex.
2. If $\bar{x}$ is the solution to the problem

$$\min_x M(x,\bar{p}); \tag{A6.1.10}$$

$$x \in Q, \tag{A6.1.11}$$

then

$$\nabla\psi(\bar{p}) = \nabla_p M(\bar{x},\bar{p}) = f(\bar{x}). \tag{A6.1.12}$$

[Here $\nabla_p M(\bar{x},\bar{p})$ is the vector of partial derivatives (with respect to p) of $M(x,p)$ evaluated at the point $(\bar{x},\bar{p})$ and $f(\bar{x})$ denotes the vector $f(\bar{x}) = (f_s(\bar{x})\ s \in S)$.]

If $c(\bar{x})$ is twice differentiable we have

$$\nabla^2\psi(\bar{p}) = f'(\bar{x})[\nabla^2 M_{xx}(\bar{x},\bar{p})]^{-1}[f'(\bar{x})]' \tag{A6.1.13}$$

Here $f'(\bar{x})$ is the vector $\{f_s\ (x) | s \in S\}$, the vector $[f'(\bar{x})]'$ is the transposed vector and $\nabla^2 M_{xx}(\bar{x},\bar{p})$ is the matrix of second derivatives (Hessian) with respect to x of function $M(\bar{x},\bar{p})$ evaluated at the point $(\bar{x},\bar{p})$.

3. The gradient of $\psi(p)$ satisfies the Lipschitz condition with the constant $1/q$:

$$\|\nabla\psi(p_1) - \nabla\psi(p_2)\| \leq \frac{1}{q}\|p_1 - p_2\|. \tag{A6.1.14}$$

4. $M(x,p)$ is stable in p in the following sense. Let X^* denote the set of solutions of the primal problem (A6.1.1)-(A6.1.3) and P^* the set of dual solutions. Then for any $p^* \in P^*$ the set of solutions of the problem

$$\min_x M(x,p^*); \tag{A6.1.15}$$

$$x \in Q, \tag{A6.1.16}$$

coincides with x^*.

Proof: The proof of the proposition clearly follows from results obtained by Polyak and Tretyakov (1973). Thus if $M(x,p)$ is stable in p, then to solve (A6.1.1)-(A6.1.3) it is sufficient to find some p^* that maximizes $\psi(p)$ and then solve (A6.1.15), (A6.1.16) using p^*. Hence the solution of (A6.1.1)-(A6.1.3) reduces to the maximization of $\psi(p)$.

Note that although $\psi(p)$ is not known explicitly, the properties of $\psi(p)$ described above give us all the information necessary to find the maximum of the function. Indeed, property 1 guarantees the convergence of many solution methods; property 2 makes it possible to use gradient, Newton, and quasi-Newton maximization methods (Chapter 1) to find the maximum of $\psi(p)$; whereas property 3 ensures the convergence of gradient methods with fixed step sizes, that is, methods in which

$$p^{\sigma+1} = p^{\sigma} + \alpha\nabla\psi(p^{\sigma}), \tag{A6.1.17}$$

provided that $\alpha < 2q$.

Solving the aggregate function can be viewed in terms of the dual function. Let x^{σ} denote the primal solution of (A6.1.4)-(A6.1.6) and λ^{σ} the corresponding dual solution. The following lemma is basic to any understanding of the essential features of iterative aggregation algorithms with constraint aggregation.

LEMMA A6.1.1 For any λ_i $(i = 1, \ldots, I)$ it is true that

$$\psi(p^{\sigma-1} + \Sigma_i\ \lambda_i^{\sigma}\theta_i^{\sigma}) \geqq \psi(p^{\sigma-1} + \Sigma_i\ \lambda_i\theta_i^{\sigma}). \tag{A6.1.18}$$

Proof: Since x^{σ}, λ^{σ} are the primal and dual solutions of (A6.1.4)-(A6.1.6), respectively, it follows that $(x^{\sigma},\lambda^{\sigma})$ is the saddle point of the Lagrangian function for this problem. Hence λ^{σ} maximizes the dual function (with respect to the classical Lagrangian) $\phi(p)$ of (A6.1.4)-(A6.1.6):

$$\phi(\lambda^{\sigma}) \geqq \phi(\lambda) \tag{A6.1.19}$$

for any λ. But

$$\phi(\lambda) = \psi(p^{\sigma-1} + \Sigma_i\ \lambda_i\theta_i^{\sigma}). \tag{A6.1.20}$$

Then the lemma follows directly from (A6.1.19) and (A6.1.20).

Lemma A6.1.1 states that (1) the aggregate problem (A6.1.4)-(A6.1.6) is equivalent to maximizing the dual function $\psi(p)$ over the linear manifold $P = \{p^{\sigma-1} + \Sigma_i\ \lambda_i\theta_i^{\sigma}\}$ and that (2) the vector

$$\tilde{p}^{\sigma} = p^{\sigma-1} + \Sigma_i\ \lambda_i^{\sigma}\theta_i^{\sigma} \tag{A6.1.21}$$

is a vector where the maximum is obtained.

In the algorithm of Martines-Soler and in that of Vakhutinsky and Dudkin, aggregation is performed over the constraint set S that has been partitioned into some disjoint subsets S_i, i = 0, 1, ..., I. The vectors θ_i^σ are assumed to be equal to the vectors $p_i^{\sigma-1} = (p_i^{\sigma-1}, s \in S_i)$. The resulting problem (A6.1.4)-(A6.1.6) is equivalent to the problem

$$\min \left\{ c(x) + \sum_{s \in S_0} p^{\sigma-1} f_s(x) + \sum_s \frac{q}{2} f_s^2(x) \right\}; \tag{A6.1.22}$$

$$\sum_{s \in S_i} p_s^{\sigma-1} f_s(x) = 0, \qquad i = 1, \ldots, I; \tag{A6.1.23}$$

$$x \in Q. \tag{A6.1.24}$$

(In the algorithm of Martines-Soler the set S_0 is always empty.) Therefore, it follows from Lemma A6.1.1 that in these algorithms the solution of the aggregate problem is equivalent to the maximization of $\psi(p)$ over the cone $(p_s = p_s^{\sigma-1}, s \in S_0, p_s = \lambda_i p_s^{\sigma-1}, s \in S_i, i = 1, \ldots, I)$.

This corresponds to the approach that views the aggregated variables as factors for the proportional change of previous approximations. For general linear dual aggregation using the vector $p^{\sigma-1}$ and the VDIs θ_i^σ, i = 1, ..., I, the lemma states that the solution of the aggregate problem is equivalent to the maximization of the dual function $\psi(p)$ on the manifold $P = \{p^{\sigma-1} + \Sigma_i \lambda_i \theta_i^\sigma\}$. Thus the vector $\tilde{p}^\sigma$, which is equal to $p^{\sigma-1} + \Sigma_i \lambda_i^\sigma \theta_i^\sigma$, where λ^σ is the dual solution of the aggregate problem at iteration σ, realizes the maximum of $\psi(p)$ on the manifold $P = \{p^{\sigma-1} + \Sigma_i \lambda_i \theta_i^\sigma\}$.

In the one-sided algorithms of Martines-Soler and of Vakhutinsky and Dudkin the dual function $\psi(p)$ is maximized over a cone; the structure of this cone makes it clear that to derive p^σ from $\tilde{p}^\sigma$ in these algorithms it is necessary to perform one iteration of some other algorithm on $\tilde{p}^\sigma$. Otherwise, the algorithm would come to a stop. That is, if we take $p^\sigma = \tilde{p}^\sigma$, the vector $p^{\sigma+1}$ would be equal to p^σ. In these algorithms p^σ is computed by the equation

$$p^\sigma = \tilde{p}^\sigma + \alpha^\sigma f(x^\sigma), \tag{A6.1.25}$$

where x^σ is the primal solution of the aggregate problem at iteration σ. In the algorithm of Martines-Soler α^σ is subject to the following constraints:

$$\alpha^\sigma \to 0, \qquad \Sigma_\sigma \alpha^\sigma = \infty. \tag{A6.1.26}$$

In the algorithm of Vakhutinsky and Dudkin α^σ is assumed to be constant for each iteration:

$$p^\sigma = \tilde{p}^\sigma + \alpha f(x^\sigma). \tag{A6.1.27}$$

The value of α is viewed as a parameter of the process that has to be selected during the computational process itself.

The aggregate problem has one other property which is very important for constructing iterative aggregation algorithms. The reason for using the formulas (A6.1.25)-(A6.1.27) to derive p^σ from $\tilde{p}^\sigma$ is clarified by this property.

LEMMA A6.1.2 Let x^σ, λ^σ be the primal and dual solutions of the aggregate problem (A6.1.4)-(A6.1.6). Then

$$\nabla\psi(p^{\sigma-1} + \Sigma_i \lambda_i^\sigma \theta_i^\sigma) = f(x^\sigma). \tag{A6.1.28}$$

Proof: We have to prove that $f(x^\sigma)$ is the gradient of $\psi(p)$ at the point $\tilde{p}^\sigma$. From the properties of the Lagrangian function we obtain that the vector x^σ is the solution to the problem

$$\min_x \left\{c(x) + \sum_{s\in S} \tilde{p}_s^\sigma f_s(x) + \sum_{s\in S} \frac{q}{2} g_s^2(x)\right\}; \qquad x \in Q. \tag{A6.1.29}$$

From the second property of modified Lagrangian functions it follows that

$$\nabla\psi(\tilde{p}^\sigma) = f(x^\sigma),$$

which completes the proof of the lemma.

From this lemma we can give the following interpretation of the one-sided algorithms of Martines-Soler and that of Vakhutinsky and Dudkin. The function $\psi(p)$ is being maximized on the basis of an iterative process that consists of two stages at each iteration. At the first stage $\psi(p)$ is maximized over some cone that contains $p^{\sigma-1}$. The result is the vector $\tilde{p}^{\sigma}$—the vector where $\psi(p)$ has a maximum value in that cone. At the second stage we derive p^{σ} from $\tilde{p}^{\sigma}$ by a shift along the gradient (in the algorithm of Martines-Soler); this shift is done with the step $\alpha^{\sigma}f(x^{\sigma})$ that tends to zero. In the Vakhutinsky-Dudkin algorithm the step size is fixed at $\alpha f(x^{\sigma})$. [Note that by step here we mean the quantities α^{σ} and α and not $\alpha^{\sigma}f(x^{\sigma})$ and $\alpha f(x^{\sigma})$, both of which should tend to zero.]

We now prove the convergence of the more general algorithms for the case where the set P^* of the vectors p^* that maximize $\psi(p)$ is bounded. We shall assume that the aggregate problem is generated using general linear dual aggregation, or equivalently that the aggregate problem is given by (A6.1.4)-(A6.1.6) and not by its limiting case (A6.1.22)-(A6.1.24). In other words, we assume that $\tilde{p}^{\sigma}$ realizes the maximum of the dual function in an arbitrary linear manifold that passes through $p^{\sigma-1}$.

Thus the generalized versions of the algorithms of Martines-Soler and of Vakhutinsky-Dudkin are distinguished by the fact that they use a more general aggregate problem, while the derivation of p^{σ} from $\tilde{p}^{\sigma}$ is as in the previous algorithms.

THEOREM A6.1.1 The algorithms (A6.1.4)-(A6.1.6), (A6.1.21), (A6.1.25), (A6.1.26); and (A6.1.4)-(A6.1.6), (A6.1.21), (A6.1.27) with $\alpha < 2q$ converge when P^* is bounded, $c(x)$ concave and $f_s(x)$ linear, to the solution under any choice of the vectors θ_i^{σ} and for any initial approximation.

Proof: Under the assumptions on $c(x)$ and $f(x)$ the global convergence follows from the well-known properties of gradient processes applied to convex functions that satisfy a Lipschitz condition. Straightforward calculations show that if

$$\|\nabla\psi(p + \alpha h) - \nabla\psi(p)\| < \frac{1}{q}\|\alpha h\|, \tag{A6.1.30}$$

where $h = \nabla\psi(p)$ and $\alpha < 2q$, then

$$\psi(p + \alpha h) - \psi(p) \geqq \alpha\left(1 - \frac{\alpha}{2q}\right)\|h\|. \tag{A6.1.31}$$

Namely, using the inequality

$$\psi(p + \alpha h) - \psi(p) = \int_0^1 (\nabla\psi(p + \tau\alpha h), \alpha h)\ d\tau, \tag{A6.1.32}$$

we obtain

$$\begin{aligned}\psi(p + \alpha h) - \psi(p) &= \alpha \int_0^1 (\nabla\psi(p), h)\ d\tau \\ &\quad + \alpha \int_0^1 (\nabla\psi(p + \tau\alpha h) - \nabla\psi(p), h)\ d\tau \\ &= \alpha\|\nabla\psi(p)\|^2 + \alpha \int_0^1 (\nabla\psi(p + \tau\alpha h) \\ &\quad - \nabla\psi(p), h)\ d\tau.\end{aligned} \tag{A6.1.33}$$

Here we made use of the fact that $h = \nabla\psi(p)$. Hence

$$\psi(p + \alpha h) - \psi(p) \geq \alpha\|\psi(p)\|^2 - \alpha \int_0^1 \|\psi(p + \tau\alpha h) - \psi(p)\|\cdot\|h\|\ d\tau.$$

Utilizing the inequality (A6.1.30), we have

$$\begin{aligned}\psi(p + \alpha h) - \psi(p) &\geq \alpha\|\nabla\psi(p)\|^2 - \frac{\alpha}{q}\int_0^1 \tau\|\alpha h\|\cdot\|h\|\ d\tau \\ &= \alpha\|\nabla\psi(p)\|^2 - \frac{\alpha^2}{q}\|\nabla\psi(p)\|^2 \int_0^1 \tau\ d\tau \\ &= \alpha\|\nabla\psi(p)\|^2 - \frac{\alpha^2}{2q}\|\nabla\psi(p)\|^2 \\ &= \alpha\left(1 - \frac{\alpha}{2q}\right)\|\nabla\psi(p)\|^2.\end{aligned} \tag{A6.1.34}$$

But $\alpha > 0$, $\|\nabla\psi(p)\| > 0$, and since $\alpha < 2q$, $(1 - \alpha/2q)$ is also greater than 0. Hence, given our assumptions,

$$\psi(p + \alpha h) \geqq \psi(p).$$

Now let us turn to the algorithms. According to the way $\tilde{p}^{\sigma}$ has been constructed, we have

$$\psi(\tilde{p}^\sigma) \geqq \psi(p^{\sigma-1}).$$

The inequality (A6.1.31) in turn guarantees that

$$\psi(p^\sigma) \geqq \psi(\tilde{p}^\sigma) + \alpha^\sigma\left(1 - \frac{\alpha^\sigma}{2q}\right)\|\nabla\psi(\tilde{p}^\sigma)\|^2$$

$$\geqq \psi(p^{\sigma-1}) + \alpha^\sigma\left(1 - \frac{\alpha^\sigma}{2q}\right)\|\nabla\psi(\tilde{p}^\sigma)\|^2 \qquad \text{(A6.1.35)}$$

In the algorithm of Vakhutinsky and Dudkin we stipulated that $\alpha < 2q$; in the algorithm of Martines-Soler the inequality $\alpha^\sigma < \beta < 2q$ for some β follows from (A6.1.25).

We now show the convergence of the algorithms to the set of dual optimal solutions P* using the inequality (A6.1.35), the fact that P* is bounded, and the fact that $\psi(p)$ is bounded from above. From (A6.1.35) for $\sigma > N$ we have

$$\psi(p^\sigma) \geqq \psi(p^N) + \sum_{\tau=N+1}^{\sigma} \alpha^\tau\left(1 - \frac{\alpha^\tau}{2q}\right)\|\psi(\tilde{p}^\tau)\|^2. \qquad \text{(A6.1.36)}$$

Let N be such that for $\tau > N$ all numbers $1 - (\alpha^\tau/2q)$ are greater than some positive number β. In the algorithm of Martines-Soler the existence of such an N follows from (A6.1.26), whereas in the Vakhutinsky-Dudkin algorithm N can be any integer.

Suppose that the quantity $\|\nabla\psi(\tilde{p}^\tau)\|^2$ for all τ be greater than some $\gamma > 0$. Then from (A6.1.36) we have

$$\psi(p^\sigma) \geqq \psi(p^N) + \beta\gamma \sum_{\tau=N+1}^{\sigma} \alpha^\tau.$$

Since according to (A6.1.26) the series $\Sigma_{\tau>N}\, \alpha^\tau$ diverges, $\psi(p^\sigma)$ goes to infinity; but this is a contradiction because $\psi(p)$ is bounded from above. Thus there exists a sequence of positive integers $N_1 = \{n_1, n_2, \ldots\}$ such that

$$\lim_{\tau\to\infty} \|\nabla\psi(p^n)\| = 0. \qquad \text{(A6.1.37)}$$

Moreover, since P* is bounded and $\psi(p)$ is convex, any level set of the type $\{p \mid \psi(p) \geqq c\}$ is also convex. Then it follows from the inequality (A6.1.35) that for all σ we have

$$\|p^* - p^{\sigma}\| < R, \tag{A6.1.38}$$

where R is a sufficiently large number and p* is an arbitrary vector in P*.

From the fact that $\psi(p)$ is convex we have

$$\|\nabla\psi(p^{\sigma}), p^* - p^{\sigma}\| \geqq \psi(p^*) - \psi(p^{\sigma}). \tag{A6.1.39}$$

Since

$$\|\nabla\psi(p^{\sigma}), p^* - p^{\sigma}\| \leqq \|\psi(p^{\sigma})\|\cdot\|p^* - p^{\sigma}\|,$$

we have

$$\|\nabla\psi(p^{\sigma})\|\cdot\|p^* - p^{\sigma}\| \geqq \psi(p^*) - \psi(p^{\sigma}). \tag{A6.1.40}$$

From (A6.1.37) and (A6.1.38) it follows that for any arbitrarily small $\varepsilon > 0$ there exists an n_{τ} in N_1 satisfying

$$\varepsilon > \|\nabla\psi(p^{n})\|\cdot\|p^* - p^{n}\| \geqq \psi(p^*) - \psi(p^{n_{\tau}}) \tag{A6.1.41}$$

is valid. Since the sequence $\psi(p^{\sigma})$ is monotone, we obtain

$$\lim_{\sigma\to\infty} \psi(p^{\sigma}) = \psi(p^*). \tag{A6.1.42}$$

The convergence of the general one-sided algorithms of Martines-Soler and of Vakhutinsky-Dudkin, that is the validity for these algorithms of the equality

$$\lim \|p^{\sigma} - P^*\| = 0, \tag{A6.1.43}$$

where $\|p^{\sigma} - P^*\| = \inf_{p^*\in P^*}\|p^{\sigma} - p^*\|$, follows from (A6.1.42) and the fact that P* is bounded. This completes the proof of the theorem.

Thus we have proven the convergence of the general one-sided algorithms of Martines-Soler and Vakhutinsky-Dudkin for the case where P* is bounded. If, however, P* is unbounded, even convergence in the functional cannot be guaranteed for the general case.

Let us note that Lemmas A6.1.1 and A6.1.2 provide us with all the information needed to construct a gradient method for maximizing $\psi(p)$ with complete directional maximization. These lemmas enable us not only to analyze previously developed iterative aggregation algorithms, but also to construct new algorithms that are computationally more efficient. Namely, we can choose some arbitrary algorithm for finding the unconstrained maximization of $\psi(p)$ and then use it as the basis for constructing an iterative aggregation algorithm.

In keeping with these remarks we now use iterative aggregation to develop a gradient algorithm with full one-dimensional maximization. Let

$$p^o = (p_s^o | s \in S), \qquad \theta^1 = (\theta_s^1 | s \in S)$$

be arbitrary initial approximate vectors. Assume that at iteration σ we have the vectors $p^{\sigma-1} = (p_s^{\sigma-1} | s \in S)$ and $\theta^\sigma = (\theta_s^\sigma | s \in S)$; we then construct and solve the following aggregate problem:

$$\min_x \left\{ c(x) + \sum_{s \in S} p_s^{\sigma-1} f_s(x) + \sum_{s \in S} \frac{q}{2} f_s^2(x) \right\}; \tag{A6.1.44}$$

$$\sum_{s \in S} \theta_s^\sigma f_s(x) = 0; \tag{A6.1.45}$$

$$x \in Q. \tag{A6.1.46}$$

Let x^σ be the solution of the problem and λ^σ the price for the constraint (A6.1.45). Then we can use the vector $p^{\sigma-1} + \lambda^\sigma \theta^\sigma$ to estimate p^σ and assume that the vector $\theta^{\sigma+1}$ is equal to $(f_s(\tilde{x}) | s \in S)$. The process is then repeated until the magnitude of the norm of the vector θ^σ becomes less than a prespecified value.

Since the algorithm in question is a gradient maximization algorithm for finding the maximum of $\psi(p)$ with full one-dimensional maximization, then given our assumptions on $c(x)$ and $f_s(x)$, $s \in S$, the algorithm will converge globally for any initial approximation.

Let us analyze this algorithm in greater detail. We can see that it is quite close to the method of multipliers (see Section 1.3). Actually, the only difference between the two methods is that in the aggregate method we solve (A6.1.44), (A6.1.46) with the additional constraint (A6.1.45). It is well known that the method of multipliers (see, e.g., Tretyakov, 1973) represents a gradient maximization method for $\psi(p)$ with a fixed step size. Thus the present algorithm is somewhat more complex due to the constraint (A6.1.45) and leads to a gradient method with full one-dimensional optimization. It seems natural to use (A6.1.44)-(A6.1.46) in place of the method of multipliers only when the additional constraint (A6.1.45) does not entail a much greater computational cost to solve the problem; therefore, the aggregate algorithm is recommended when all the constraints of the original problem are linear. In that case the solution cost of (A6.1.44), (A6.1.46) is practically the same as that of the problem with the additional constraint (A6.1.45).

It should be noted that several authors have commented on the fairly slow rate of convergence (with respect to the dual variables) of the method of multipliers (see Fedorenko, 1977). In our opinion, this slow rate of convergence stems mainly from the fact that the method of multipliers maximizes $\psi(p)$ using a fixed-step gradient algorithm. Recently, attempts have been made to improve the rate of convergence by constructing Newton and quasi-Newton variants of those algorithms (see, e.g., Fletcher, 1973). However, even with these new algorithms the desire to reduce the solution of the original problem to that of solving a sequence of unconstrained optimization problems necessitates the use of Newton or quasi-Newton algorithms with a fixed step size.

It follows from the discussion above that even when we deal with Newton and quasi-Newton algorithms the addition of a single constraint of the form (A6.1.45) results in algorithms with full one-dimensional maximization. However, these algorithms still have the usual difficulty with large-scale problems, of having to store a very large matrix. Therefore, in our opinion the method of conjugate gradients for maximizing $\psi(p)$ is of considerable interest. It is distinguished from the gradient algorithm described above only by the fact that $\theta^{\sigma+1}$ is no longer set equal to the vector $f(x^{\sigma})$ but rather to the vector

$$\theta^{\sigma+1} = f(x^\sigma) + \frac{\|f(x^\sigma)\|^2}{\|f(x^{\sigma-1})\|^2}\,\theta^\sigma. \tag{A6.1.47}$$

As usual, the computational cost of the resulting algorithm is practically equivalent to that of the gradient method, but it possesses a considerably faster rate of convergence.

Thus we can see that by using Lemmas A6.1.1 and A6.1.2 we can realize practically any algorithm that maximizes $\psi(p)$ on the basis of the aggregate problem (A6.1.4)-(A6.1.6). In addition, the convergence of the corresponding algorithms follows from the available results concerning the convergence of the relevant unconstrained algorithm. Thus in realizing the sequential group coordinate descent algorithm for maximizing $\psi(p)$, we obtain the following one-level algorithm. Let the set of constraints S be divided into I not necessarily disjoint subsets S_i $(S = \cup_{i=1}^{I} S_i)$. Each iteration of the algorithm includes the solution of I problems sequentially. The ith problem at iteration σ will be

$$\min_x \left\{ c(x) + \sum_{s\in\bar{S}_i} p_s^\sigma f_s(x) + \sum_{s\in\bar{\bar{S}}_i} p_s^{\sigma-1} f_s(x) + \sum_{s\in S} \frac{q}{2} f_s^2(x) \right\}; \tag{A6.1.48}$$

$$\bar{S}_i = \left(\bigcup_{j=1}^{i-1} S_j\right) \setminus S_i, \qquad \bar{\bar{S}}_i = S \setminus \bigcup_{j=1}^{i} S_j; \tag{A6.1.49}$$

$$f_s(x) = 0, \qquad s \in S_i; \qquad x \in Q. \tag{A6.1.50}$$

Let p_s^σ, $s \in S_i$, denote the price of the constraints (A6.1.49); in addition if $s \in S_i \cap S_j$ and $i > j$, let p_s^σ first denote the price of the constraint determined in problem j until the problem is solved, and after that according to problem i.

Finally, we can try to maximize $\psi(p)$ using the iterative aggregation algorithm described in Chapter 5. In solving (A6.1.4)-(A6.1.6) we maximize $\psi(p)$ over the linear manifold that passes through $p^{\sigma-1}$ and is spanned by the vectors θ_i^σ, $i = 1, \ldots, I$, along the manifold.

This is precisely the way we generate the general iterative aggregation algorithms with constraint aggregation that are described in Chapter 6. The convergence of the resulting algorithms follows directly from the convergence of the corresponding algorithms for the solution of unconstrained maximization problems.

In conclusion we note that the aggregation of increments of dual variables retains clear-cut economic interpretation of the original aggregation methods and does not increase the cost of constructing and solving the aggregate problem when compared with aggregation of the dual variables themselves. On the other hand, the use of these algorithms signifies a passage to considerably more effective algorithms for maximizing $\psi(p)$, which accounts for a much faster rate of convergence in the resulting algorithms. Thus we can see that our approach to constraint aggregation enables us to construct a variety of algorithms for the solution of constrained optimization problems. Let us present a general theorem concerning the convergence of these algorithms.

THEOREM A6.1.2 Let the algorithm for maximizing convex functions $g(p)$ which maps p^σ into $A(p^{\sigma-1})$ be defined in $|S|$-dimensional Euclidean space. In addition, suppose that

(i) $A(p^{\sigma-1}) \in \operatorname{Arg} g(p^{\sigma-1} + \Sigma_i \lambda_i \theta_i^\sigma)$ (A6.1.51)

(here $\operatorname{Arg} g(p^{\sigma-1} + \Sigma_i \lambda_i \theta_i^\sigma)$ is the set of maxima of $g(p)$ in the linear manifold that passes through $p^{\sigma-1}$ and is spanned by the vectors θ_i^σ, $i = 1, \ldots, I$);

(ii) $\theta^{\sigma+1}$ is expressed as some function of $\{\theta^1\}, \ldots, \{\theta^\sigma\}, \nabla g(p^0), \ldots, \nabla g(p^\sigma), p^0, p^1, \ldots, p^\sigma$:

$$\theta_i^{\sigma+1} = F(\{\theta^1\}, \ldots, \{\theta^\sigma\}, \nabla g(p^\sigma), \ldots, \nabla g(p^0), p^0, \ldots, p^\sigma) \quad i = 1, \ldots, I^{\sigma+1} \tag{A6.1.52}$$

where $\{\theta^k\}$ is the collection of vectors θ_i^k, $i = 1, \ldots, I^k$.

Then if this algorithm globally (locally) converges in the case of arbitrary concave functions, the algorithm (A6.1.4)-(A6.1.6),

(A6.2.1) for the problems (A6.1.1)-(A6.1.3) with $c(x)$ concave, $f_s(x)$ linear and Q a convex set also converges globally (locally).

We assume that all algorithms take

$$\theta_i^{\sigma+1} = F(\{\theta^1\}, \ldots, f(x^1), \ldots, f(x^\sigma), \ldots, p^0, \ldots, p^\sigma), \tag{A6.1.53}$$

where $f(x^k) = \{f_s(x^k) | s \in S\}$ and p^σ is assumed to be equal to $\tilde{p}^\sigma$.

Proof: The proof of the theorem follows directly from Proposition A6.1.1 and Lemmas A6.1.1 and A6.1.2.

A straightforward corollary of this result is that given the same assumptions the algorithm (6.2.27)-(6.2.31) globally converges. Let us also note that when (6.1.1)-(6.1.3) is a linear programming problem, the algorithm (6.2.27)-(6.2.30), (6.2.32) will be finite. This follows from the results of Polyak and Tretyakov (1973).

APPENDIX 6.2 MATHEMATICAL THEORY OF THE TWO-LEVEL ALGORITHM WITH BOTH CONSTRAINTS AND VARIABLES AGGREGATION

In this appendix we prove that all stationary points of the algorithm (6.3.10)-(6.3.25) are optimal for a problem of convex programming and then describe a special case of the algorithm which we prove converges locally for a general convex programming problem. The local convergence theorem is stated without proof. The proof of convergence for linear programming with simple aggregation where the aggregate problem solved in the algorithm differs somewhat from the aggregate problem used in this chapter can be found in Vakhutinsky, Dudkin, and Rivkin (1979). The proof of local convergence for the iterative aggregation algorithm using simple aggregation to solve the general nonlinear programming problem with nondegenerate optimum is presented in Vakhutinsky (1979).

THEOREM A6.2.1 Assume that the function $c(x)$ in (6.3.1)-(6.3.4) is convex and the functions $f_s(x)$, $h(x)$ are concave. Further, suppose that all the functions are continuously differentiable in the neighborhood of the positive octant. Then:

1. Any stationary point x^* of the general two-level algorithm (6.3.10)-(6.3.25) is also the optimal solution of the original problem (6.3.1)-(6.3.4).
2. The following relations hold at the stationary point:

$$z_r^* = e_r^* \text{ for all } r = 1, \ldots, R;$$
$$v_{sj} = 0 \text{ for all } s \in S,\ j = 1, \ldots, J.$$
$$\text{If } u_s > 0 \text{ for } s \in S_0, \text{ then } p_s = 0;$$
$$\text{if } u_s = 0 \text{ for } s \in S_0, \text{ then } f_s(x^*) = 0.$$

Note: From the theorem it follows that if the process converges at all, it converges to the solution of the original problem since the limit point will always be a stationary point.

Proof: Let x_j^* $(j = 1, \ldots, J)$, p_s^* $(s \in S)$, p_t^* $(t \in L_j,\ j = 1, \ldots, J)$ be the components of the stationary point. The fact that the point is a stationary point means that if $x_j^{\sigma-1} = x_j^*$, $p_s^{\sigma-1} = p_s^*$, $p_t^{\sigma-1} = p_t^*$, then $x_j^{\sigma} = x_j^*$, $p_s^{\sigma} = p_s^*$, $p_t^{\sigma} = p_t^*$, where each sub ranges over a corresponding set. From the formulas (6.3.27) we also obtain that $\hat{x}_j^{\sigma} = x_j^*$, $\hat{p}_s^{\sigma} = p_s^*$, $\hat{p}_t^{\sigma} = p_t^*$. Subsequently, we shall use these equalities without citing a specific reference.

We use * to denote the aggregate indices and functions that are computed using the stationary point. Note that since $c(x_1,\ldots,x_J)$, $f_s(\cdot)$ and $h_t(x_j)$ are convex and continuously differentiable in a neighborhood of the positive octant, the same hold for $c^*(z)$, $f_t^*(z)$, $\hat{f}_s^*(z)$.

Let the primal and dual solutions of the aggregate problem (6.3.13)-(6.3.15) at the stationary point be denoted by z^*, u_s^* $(s \in S_0)$, λ_i $(i = 1, \ldots, I)$, respectively. From the Karush-Kuhn-Tucker optimization criteria we obtain the following relations for the aggregate problem:

$$\frac{\partial c^*(z^*)}{\partial z_r} - \sum_{i=0}^{I} \lambda_i^* \frac{\partial f_i^*(z^*)}{\partial z_r} - Q_0'\beta_r^*(z_r^* - e_r^*) - Q_0 \sum_{s\in S_0} \sum_{r=1}^{R} b_{sr}^*$$
$$\times \left[\sum_{r'=1}^{R} b_{sr'}^*(z_{r'}^* - e_{r'}^*) + \hat{f}_s^*(e^*) + u_s^* \right] = \begin{cases} 0 & \text{if } z_r^* > 0, \\ \le 0 & \text{if } z_r^* = 0, \end{cases}$$
$$r = 1, \ldots, R. \tag{A6.2.1}$$

$$-Q_0\left[\sum_{r=1}^{R} b^*_{sr}(z^*_r - e^*_r) + f^*_s(e^*) + u^*_s\right] - p^*_s$$
$$= \begin{cases} 0 & \text{if } u^*_s > 0, \\ \le 0 & \text{if } u^*_s = 0, \end{cases} \qquad s \in S_0, \tag{A6.2.2}$$

where to preserve the uniformity of notation we set λ^*_0 equal to 1. From (6.3.10)-(6.3.12) it follows that

$$\frac{\partial c^*(z^*)}{\partial z_r} = \sum_{k\in K}\left[\frac{\partial c(\Phi^* z^*)}{\partial \xi_k} - \sum_{j=1}^{J}\sum_{t\in L_j} p^*_t \frac{\partial h_t(\Phi^* z^*)}{\partial \xi_k}\right]\varphi^*_{kr}, \tag{A6.2.3}$$

$r = 1, \ldots, R.$

$$\frac{\partial f^*_i(z^*)}{\partial z_r} = \sum_{k\in K} \varphi^*_{kr} \sum_{s\in S_i} p^*_s \frac{\partial f_s(\Phi^* z^*)}{\partial \xi_k}, \quad i = 0, \ldots, I. \tag{A6.2.4}$$

Substituting (A6.2.3), (A6.2.4) into (A6.2.1), we obtain

$$\sum_{j=1}^{J}\sum_{k\in K_j}\left[\frac{\partial c(\Phi^* z^*)}{\partial \xi_k} - \sum_{t\in L_j} p^*_t \frac{\partial h_t(\Phi^* z^*)}{\partial \xi_k} - \sum_{s\in S} \tilde{p}^*_s \frac{\partial f_s(\Phi^* z^*)}{\partial \xi_k}\right]\varphi^*_{kr}$$
$$+ Q'_0\beta^*_r(e^*_r - z^*_r) - Q_0 \sum_{s\in S_0} b^*_{sr} \sum_{r'=1}^{R} b^*_{sr'}(z^*_{r'} - e^*_{r'})$$
$$\left. + \hat{f}^*_s(e^*) + u^*_s\right] = \begin{cases} 0 & \text{if } z^*_r > 0, \\ \le 0 & \text{if } z^*_r = 0, \end{cases} \tag{A6.2.5}$$

where $\tilde{p}^*_s = \lambda^*_i p^*_s$, $s \in S_i$, $i = 0, \ldots, I$ [see (6.3.23)]. We introduce the following functions:

$$T^*_j(x_j) = c^*_j(x_j) - \sum_{s\in S} p^*_s f^*_{sj}(x_j) - \sum_{t\in L_j} p^*_t h_t(x_j); \tag{A6.2.6}$$

$$\tilde{T}^*_j(z) = T^*_j(\Phi^* z), \tag{A6.2.7}$$

where $c^*_j(x_j)$, $f^*_{sj}(x_j)$ are given by (6.3.17), (6.3.18). It is easy to see that by using (A6.2.6), (A6.2.7) we can rewrite (A6.2.5) as follows:

$$\sum_{i=1}^{J} \frac{\partial \tilde{T}_j^*(z^*)}{\partial z_r} + Q_0'\beta_r^*(e_r^* - z_r^*) - Q_0 \sum_{s\in S_0} b_{sr}^*\left[\sum_{r'=1}^{R} b_{sr'}^*(z_{r'}^* - e_{r'}^*) + f_s^*(e^*) + u_s^*\right] = \begin{cases} 0 & \text{if } z_r^* > 0, \\ \le 0 & \text{if } z_r^* = 0, \end{cases} \quad r = 1, \ldots, R. \tag{A6.2.8}$$

It is clear that each relation (A6.2.8) remains valid if we multiply its left-hand side by $(e_r^* - z_r^*)$, $r = 1, \ldots, R$, and then sum the left-hand sides over all $r = 1, \ldots, R$. As a result we obtain the inequality

$$\sum_{j=1}^{J} \sum_{r=1}^{R} \frac{\partial \tilde{T}_j^*(z_j^*)}{\partial z_r} (e_r^* - z_r^*) + Q_0' \sum_{r=1}^{R} \beta_r^*(e_r^* - z_r^*)^2 - Q_0 \sum_{s\in S_0} \left[\sum_{r=1}^{R} b_{sr}^*(e_r^* - z_r^*)\right]\left[\sum_{r=1}^{R} b_{sr}^*(z_r^* - e_r^*) + \hat{f}_s^*(e^*) + u_s^*\right] \le 0. \tag{A6.2.9}$$

Now $x_j^* = (\xi_k^* | k \in K_j)$, v_{sj}^* $(s \in S)$, p_t^* $(t \in L_j)$ is the optimal solution of the local problem (6.3.22)-(6.3.24) of subsystem j. Applying the Karush-Kuhn-Tucker optimization criteria to the local problem of subsystem j, we obtain the relation

$$\frac{\partial c_j^*(x_j^*)}{\partial \xi_k} - \sum_{s\in S} p_s^* \frac{\partial f_{sj}^*(x_j^*)}{\partial \xi_k} - \sum_{t\in L_j} p_t^* \frac{\partial h_t(x_j^*)}{\partial \xi_k} - Q \sum_{s\in S} \eta_{sk}^*\left[\sum_{k'\in K_j} \eta_{sk'}^*\left(\xi_{k'}^* - \sum_{r=1}^{R} \varphi_{k'r}^* z_r^*\right) + v_{sj}^*\right] = \begin{cases} 0 & \text{if } \xi_k^* > 0, \\ \le 0 & \text{if } \xi_k^* = 0, \end{cases} \quad k \in K_j, \tag{A6.2.10}$$

where η_{sk}^* is the coordinate of the gradient vector $g_{sj}^*(\Phi^* z^*)$, $k \in K_j$,

$$\eta_{sk}^* = \frac{\partial f_{sj}^*(\Phi_j^* z^*)}{\partial \xi_k}, \qquad k \in K_j. \tag{A6.2.11}$$

We can use the function $T_j^*(x_j)$ defined in (A6.2.6) to rewrite the relation (A6.2.10) as follows:

$$\frac{\partial T_j^*(x_j^*)}{\partial \xi_k} - Q \sum_{s \in S} \eta_{sk}^* \left[\sum_{k' \in K_j} \eta_{sk'}^* \left(\xi_{k'}^* - \sum_{r=1}^{R} \varphi_{k'r} z_r^* \right) + v_{sj}^* \right]$$

$$= \begin{cases} 0 & \text{if } \xi_k^* > 0, \\ \leq 0 & \text{if } \xi_k^* = 0, \end{cases} \qquad k \in K_j. \tag{A6.2.12}$$

Multiply both sides of (A6-2.12) by φ_{kr}^*. From (6.3.7) $\varphi_{kr}^* = 0$ when $\xi_k^* = 0$ $(r = 1, \ldots, R)$, the relations (A6.2.12) thus are transformed into exact equalities. Summing these equalities for all $k \in K_j$, we obtain

$$\sum_{k \in K_j} \varphi_{kr}^* \frac{\partial T_j^*(x_j^*)}{\partial \xi_k} - Q \sum_{s \in S} \left(\sum_{k \in K_j} \varphi_{kr}^* \eta_{sk}^* \right) \left[\sum_{k \in K_j} \eta_{sk}^* \left(\xi_k^* \right. \right.$$

$$\left. \left. - \sum_{r'=1}^{R} \varphi_{kr'}^* z_{r'}^* \right) + v_{sj}^* \right] = 0, \qquad j = 1, \ldots, J, \; r = 1, \ldots, R. \tag{A6.2.13}$$

From the fact that $x_j^* = \Phi_j^* e^*$ and from (A6.2.7), it follows that

$$\sum_{k \in K_j} \varphi_{kr}^* \frac{\partial T_j(x_j^*)}{\partial \xi_k} = \frac{\partial \tilde{T}_j^*(e^*)}{\partial z_r} \tag{A6.2.14}$$

Substituting the right-hand side of (A6.2.14) into (A6.2.13), multiplying both parts of (A6.2.13) by $(z_r^* - e_r^*)$ and then summing over $j = 1, \ldots, J$, $r = 1, \ldots, R$ and adding both sides of the resulting equality to (A6.2.9), we obtain

$$\sum_{j=1}^{J} \sum_{r=1}^{R} \left(\frac{\partial \tilde{T}_j^*(z^*)}{\partial z_r} - \frac{\partial \tilde{T}_j^*(e^*)}{\partial z_r} \right) (e_r^* - z_r^*) + Q_0' \sum_{r=1}^{R} \beta_r^* (e_r^* - z_r^*)^2$$

$$+ Q_0 \sum_{s \in S_0} \left[\sum_{r=1}^{R} b_{sr}^* (e_r^* - z_r^*) \right] \left[\sum_{r=1}^{R} b_{sr}^* (e_r^* - z_r^*) - \hat{f}_s^*(e^*) - u_s^* \right]$$

$$+ Q \sum_{j=1}^{J} \sum_{s \in S} \left[\sum_{k \in K_j} \sum_{r=1}^{R} \varphi_{kr}^* \eta_{sk}^* (e_r^* - z_r^*) \right] \left[\sum_{k \in K_j} \eta_{sk}^* \left(\xi_k^* - \sum_{r=1}^{R} \varphi_{kr}^* z_r^* \right) \right.$$

$$\left. + v_{sj}^* \right] \leqq 0. \tag{A6.2.15}$$

From the definition (A6.2.6), (A6.2.7) of $\tilde{T}_j^*(z)$ and the conditions of the theorem it follows that $\tilde{T}_j^*(z)$ is a convex function. Hence the inequality $(\nabla\tilde{T}_j^*(a) - \nabla\tilde{T}_j^*(b), a - b) < 0$ is valid, where $\nabla\tilde{T}_j^*(a)$ is the gradient of $\tilde{T}_j^*$ evaluated at a. Substituting z* and e* into this result for a and b yields

$$\sum_{r=1}^{R} \left[\frac{\partial\tilde{T}_j^*(z)}{\partial z_r} - \frac{\partial\tilde{T}_j^*(e^*)}{\partial z_r}\right] (e_r^* - z_r^*) \geqq 0. \tag{A6.2.16}$$

Now let us consider the last two terms from the left-hand side of (A6.2.15). Replacing ξ_k^* in (A6.2.15) by the right-hand side of (6.3.6) and using (6.3.7), then from (A6.2.16) and (A6.2.15) it follows that

$$\begin{aligned} &Q_0' \sum_{r=1}^{R} \beta_r^*(e_r^* - z_r^*)^2 + Q_0 \sum_{s\in S_0} \left[\sum_{r=1}^{R} b_{sr}^*(e_r^* - z_r^*)\right] \\ &\times \left[\sum_{r=1}^{R} b_{sr}^*(e_r^* - z_r^*) - f_s^*(e^*) - u_s^*\right] \\ &+ Q \sum_{j=1}^{J} \sum_{s\in S} \left[\sum_{r=1}^{R} \sum_{k\in K_j} \eta_{sk}^*\varphi_{kr}^*(e_r^* - z_r^*)\right] \\ &\times \left[\sum_{r=1}^{R} \sum_{k\in K_j} \eta_{sk}^*\varphi_{kr}^*(e_r^* - z_r^*) + v_{sj}^*\right] \leqq 0. \end{aligned} \tag{A6.2.17}$$

Consider the following subsets of the set S:

$$S_j^+ = \{s \in S | v_{sj}^* > 0\}; \qquad S_j^- = \{s \in S | v_{sj}^* = 0\};$$
$$S_0^+ = \{s \in S_0 | u_s^* > 0\}; \qquad S_0^- = \{s \in S_0 | u_s^* = 0\}.$$

From (6.3.7), (6.3.10) it follows that $f_s^*(e^*) = f_s(x^*)$ at the stationary point. The relation (6.3.24) for $s \in S_0$ assumes the following form at the stationary point:

$$p_s^* = \{p_s^* + \alpha Q f_s(x^*)\}_{(+)}, \qquad s \in S_0. \tag{A6.2.18}$$

From the formula (A6.2.2) we obtain the relations for $s \in S_0^+$ and $s \in S_0^-$:

$$u_s^* = \sum_{r=1}^{R} b_{sr}^*(e_r^* - z_r^*) - f_s(x^*) - \frac{p_s^*}{Q_0} > 0, \qquad s \in S_0^+; \tag{A6.2.19}$$

$$\sum_{r=1}^{R} b_{sr}^*(e_r^* - z_r^*) - f_s(x^*) \leqq \frac{p_s^*}{Q_0}, \qquad s \in S_0^-. \tag{A6.2.20}$$

Let us demonstrate the validity of the following inequalities:

$$Q_0\left[\sum_{r=1}^{R} b_{sr}^*(e_r^* - z_r^*)\right]\left[\sum_{r=1}^{R} b_{sr}^*(e_r^* - z_r^*) - f_s(x^*) - u_s^*\right]$$

$$\geqq \begin{cases} \dfrac{(p_s^*)^2}{Q_0} & \text{for } s \in S_0^+, \\ Q_0\left[\sum\limits_{r=1}^{R} b_{sr}^*(e_r^* - z_r^*) - f_s(x^*)\right]^2 & \text{for } s \in S_0^-. \end{cases} \tag{A6.2.21}$$

The proof of (A6.2.21) for $s \in S_0^+$ proceeds as follows. In view of (A6.2.19), the second factor in brackets on the left-hand side of (A6.2.21) is equal to p_s^*/Q_0. At the same time it follows from (A6.2.19) that

$$\sum_{r=1}^{R} b_{sr}^*(e_r^* - z_r^*) > f_s(x^*) + \frac{p_s^*}{Q_0}. \tag{A6.2.22}$$

If $p_s^* > 0$ it follows from (A6.2.18) that $f_s(x^*) = 0$, from which, in view of (A6.2.22), we obtain (A6.2.21) for $s \in S_0^+$. If, however, $p_s^* = 0$, it follows that the left-hand side of (A6.2.21) is also equal to zero and the inequality (A6.2.21) is then trivially satisfied.

To prove (A6.2.21) for $s \in S_0^-$, let us reduce the left-hand side of (A6.2.21) to the form

$$Q_0\left[\sum_{r=1}^{R} b_{sr}^*(e_r^* - z_r^*) - f_s(x^*)\right]^2 + Q_0 f_s(x^*)$$

$$\times \left[\sum_{r=1}^{R} b_{sr}^*(e_r^* - z_r^*) - f_s(x^*)\right]. \tag{A6.2.23}$$

To show that the second term in (A6.2.23) is nonnegative is tantamount to proving (A6.2.21) for $s \in S_0^-$. From (A6.2.18) it follows that $f_s(x^*) \leqq 0$. If $f_s(x^*) = 0$, then the second term in (A6.2.23) is zero. If, however, $f_s(x^*) < 0$, we obtain from (A6.2.18) that $p_s^* = 0$; and then in view of (A6.2.20) it follows that $\Sigma_{r=1}^R b_{sr}^*(e_r^* - z_r^*) - f_s(x^*) \leqq 0$. Thus in this case, too, the second term is not less than zero. Hence the inequality (A6.2.21) has been proven.

Partial maximization of (6.3.20) in relation to v_{sj} gives in particular

$$\sum_{k\in K_j} \sum_{r=1}^{R} \eta_{sk}^* \varphi_{kr}^* (e_r^* - z_r^*) + v_{sj}^* = -\frac{\tilde{p}_s^*}{Q} \quad \text{for } s \in S_j^+. \tag{A6.2.24}$$

Replacing the first term on the left-hand side of (A6.2.17) by a smaller quantity determined in (A6.2.21) and rewriting the second term according to (A6.2.24), we obtain

$$\begin{aligned} &Q_0' \sum_{\tau=1}^{R} \beta_\tau^* (e_\tau^* - z_\tau^*)^2 + \frac{1}{Q_0} \sum_{s\in S_0^+} (p_s^*)^2 \\ &+ Q_0 \sum_{s\in S_0^-} \left[\sum_{\tau=1}^{R} b_{s\tau}^* (e_\tau^* - z_\tau^*) - f_s(x^*) \right]^2 \\ &+ Q \sum_{j=1}^{J} \left\{ \sum_{s\in S_j^-} \left[\sum_{k\in K_j} \sum_{\tau=1}^{R} \eta_{sk}^* \varphi_{k\tau}^* (e_\tau^* - z_\tau^*) \right]^2 \right. \\ &\left. + \sum_{s\in S_j^+} \left[\left(\frac{\tilde{p}_s^*}{Q} \right)^2 + \frac{\tilde{p}_s^*}{Q} v_{sj}^* \right] \right\} \leq 0. \end{aligned} \tag{A6.2.25}$$

Since all the terms on the left-hand side of (A6.2.25) are nonnegative, it follows that they all must be zero. Since $\beta_r^* > 0$ for all $r = 1, \ldots, R$, we obtain

$$z_r^* = e_r^*, \qquad r = 1, \ldots, R. \tag{A6.2.26}$$

Also it follows from (A6.2.25) that

$$\tilde{p}_s^* = 0 \qquad \text{for } s \in \cup S_j^+; \tag{A6.2.27}$$

that is, $\lambda_i^* p_s^* = 0$ for $s \in S_i$, $s \in \cup S_j^+$. Let us now show that

$$\lambda_i^* p_s^* = p_s^*, \qquad s \in S_i, \qquad i = 1, \ldots, I. \tag{A6.2.28}$$

To prove (A6.2.28), it is sufficient to establish that if for a given $i \neq 0$ there exists an $s \in S_i$ such that $p_s^* > 0$, it follows that $\lambda_i^* = 1$. Since $\hat{p}^* = p^*$ at the stationary point it follows that (6.3.24) for it has the form

$$p_s^* = [\lambda_i^* p_s^* + \alpha Q f_s(x^*)]_{(+)}, \qquad s \in S_i,\ i = 1, \ldots, I. \tag{A6.2.29}$$

Since $z_r^* = e_r^*$ for all $r = 1, \ldots, R$, we obtain from (6.3.7), (6.3.10), (6.3.11) that

$$f_i(z^*) = \sum_{s \in S_i} p_s^* f_s(x^*). \tag{A6.2.30}$$

Now consider an i for which $\lambda_i^* > 1$. If for this i all $p_s^* = 0$ ($s \in S_i$), it follows that (A6.2.28) is trivially satisfied. If, however, for some s it is true that $p_s^* > 0$, it follows from (A6.2.29) for such s that

$$p_s^*(1 - \lambda_i^*) = \alpha Q f_s(x^*).$$

Since $\lambda_i^* > 1$ and $p_s^* > 0$, it follows for $s \in S_i$ that $f_s(x^*) < 0$. From this we obtain that $\Sigma_{s \in S_i} p_s^* f_s(x^*) < 0$. In view of (A6.2.26) this means that $f_i^*(z^*) < 0$; that is, the ith aggregate constraint is not binding at the solution of the aggregate problem and hence its dual variable λ_i^* must be identically 0. This, however, contradicts the assumption that $\lambda_i^* > 1$.

Finally, assume that for some i, $\lambda_i^* < 1$ and that there exists a nonempty subset $s \in S_i$ such that $p_s^* > 0$. By identical reasoning we obtain that $f_i^*(z_i^*) > 0$, which implies that z^* is not a feasible solution of the aggregate problem, which is true by assumption. Thus equation (A6.2.28) has been proven.

In view of (A6.2.28) the relation (6.3.24) at the stationary point has the form

$$p_s^* = [p_s^* + \alpha Q f_s(x^*)]_{(+)}, \qquad s \in S. \tag{A6.2.31}$$

From (A6.2.30) we obtain that if $p_s^* > 0$, then $f_s(x^*) = 0$, and if $p_s^* = 0$, then $f_s(x^*) < 0$. We can summarize this idea as

$$f_s(x^*) = \begin{cases} 0 & \text{if } p_s^* > 0, \\ \leqq 0 & \text{if } p_s^* = 0, \end{cases} \qquad s \in S. \tag{A6.2.32}$$

Next, the following relations hold for the constraints (6.3.21) and the variables p_t ($t \in L_j$) of the jth local problem (6.3.20)-(6.3.22) that correspond to those constraints:

$$h_t(x^*) = \begin{cases} 0 & \text{if } p_t^* > 0, \\ \leqq 0 & \text{if } p_t^* = 0, \end{cases} \qquad t \in L_j,\ j = 1, \ldots, J. \tag{A6.2.33}$$

Let us show that

$$\sum_{k \in K_j} \eta_{sk}^* \left(\xi_k^* - \sum_{r=1}^{R} \varphi_{kr}^* z_r^* \right) + v_{sj}^* = 0, \qquad s \in S,\ j = 1, \ldots, J. \tag{A6.2.34}$$

If $s \in S_j^-$, then (A6.2.34) follows from (A6.2.26) and (6.3.6). If, however, $s \in S_j^+$, it follows from (A6.2.11) that the left-hand side of (A6.2.34) is equal to $-\tilde{p}_s^*/Q$, which, in turn, is equal to zero from (A6.2.27). Note that $z^* = e^*$ and $\Phi_j^* e^* = x_j^*$; hence it follows from (6.3.17) and (6.3.18) that

$$\frac{\partial c_j^*(x_j^*)}{\partial \xi_k} = \frac{\partial c(x^*)}{\partial \xi_k}, \qquad \frac{\partial f_{sj}^*(x_j^*)}{\partial \xi_k} = \frac{\partial f_s(x^*)}{\partial \xi_k}, \qquad k \in K. \tag{A6.2.35}$$

Now (A6.2.10) can be restated in terms of (A6.2.34), (A6.2.35), and (A6.2.28):

$$\frac{\partial c(x^*)}{\partial \xi_k} - \sum_{s \in S} p_s^* \frac{\partial f_s(x^*)}{\partial \xi_k} - \sum_{t \in L_j} p_t^* \frac{\partial h_t(x_j^*)}{\partial \xi_k} = \begin{cases} 0 & \text{if } \xi_k^* > 0, \\ \leq 0 & \text{if } \xi_k^* = 0. \end{cases} \tag{A6.2.36}$$

The formulas (A6.2.32), (A6.2.33), (A6.2.36) establish the necessary and sufficient conditions for the optimization criterion of the original convex programming problem (6.3.1)-(6.3.4).

To prove the second part of the theorem, note that from (A6.2.25) it follows that $p_s^* = 0$ for $s \in S_0^+$. From (A6.2.34) we obtain that

$v^*_{sj} = 0$ for all $s \in S$ (i.e., $S^-_j = S$ and $S^+_j = \emptyset$ for all $j = 1, \ldots, J$). Finally, the equality $f_s(x^*) = 0$ for $s \in S^-_0$ follows from the fact that from (A6.2.25) it follows that $\Sigma^R_{r=1} b^*_{sr}(e^*_r - z^*_r) - f_s(x^*) = 0$ for $s \in S^-_0$, and that $z^*_r = e^*_r$. This completes the proof of Theorem A6.2.1.

Next we study the local convergence of the algorithm (6.3.10)-(6.3.25) subject to the following constraints:

(a) The local constraints (6.3.2) are either absent or are included in the global constraints (6.3.1).

(b) The set S_0 coincides with the entire set S; the sets $S_i = \emptyset$ for $i \neq 0$ and $Q'_0 = Q_0$.

(c) The coefficients β^σ_r, φ^σ_{kr}, e^σ_r are continuous functions of $x^{\sigma-1}$ satisfying

$$\varphi^\sigma_{kr} = \varphi_{kr}(x^{\sigma-1}) \geqq 0,$$

$$e^\sigma_r = e_r(x^{\sigma-1}) \geqq 0 \qquad \text{[when (6.33) holds]},$$

$$\beta^\sigma_r = \beta_r(x^{\sigma-1}) \qquad \text{where } \beta_r(x) > \beta_0 > 0.$$

(d) All functions in the objective function and constraints of the original problem are twice continuously differentiable.

Thus to solve the original problem (6.3.1), (6.3.3), (6.3.4) we deal with a limiting case of the process (6.3.10)-(6.3.25). We introduce the following notation to facilitate the presentation of the theorem. Let x^*, p^* denote the primal and dual solutions of the original problem. Let us consider the set S^* of constraints that are binding at the optimal point x^*:

$$S^* = \{s \in S \mid f_s(x^*) = 0\}$$

and also consider the set of positive coordinates of the vector

$$K^* = \{k \in \{1, \ldots, K\} \mid \xi^*_k > 0\}.$$

By analogy with linear programming we will refer to S^* and K^* as the set of basic constraints and the basic solution set, respectively.

Let G denote the "truncated" $|S^*| \times |K^*|$-dimensional matrix obtained from the Jacobian matrix at x^* of the system of functions in (6.3.1) by eliminating all columns and rows not belonging to K^* and all rows not belonging to S^*, that is,

$$G = \frac{\partial f_s(x^*)}{\partial \xi_k}, \qquad s \in S^*, \qquad k \in K^*. \tag{A6.2.37}$$

Let $L(x,p)$ denote the Lagrangian function:

$$L(x,p) = c(x) - \sum_{s \in S} p_s f_s(x).$$

Similarly, let D denote the reduced $|K^*|$-dimensional square matrix derived from the matrix of second derivatives in ξ_k of $L(x,p)$ evaluated at the point x^*, that is,

$$D = \frac{\partial^2 L(x^*,p^*)}{\partial \xi_k \, \partial \xi_{k'}}, \qquad k,\ k' \in K^*. \tag{A6.2.38}$$

We assume that the following second-order stability conditions are valid at the local maximum x^*, p^*:

(A) The rows of the matrix G are linearly independent.

(B) The matrix D is negative definite on the subspace $N = \{\xi | G\xi = 0\}$; that is, if $G\xi = 0$, $\xi \neq 0$, then

$$(D\xi, \xi) < 0. \tag{A6.2.39}$$

(C) Strong complementary is satisfied at the maximum, that is,

$$p_s^* > 0 \text{ for } s \in S^*, \qquad \frac{\partial L(x^*,p)}{\partial \xi_k} < 0 \text{ for } k \in K^*. \tag{A6.2.40}$$

The conditions (A)-(C) are well known (see, e.g., Fiacco and McCormick, 1968).

For the linear programming problems it follows from these conditions that the solution is unique and nondegenerate for the following reasons. From condition (B) it follows that the subspace $N = 0$ since in the linear programming problem the $|K^*| \times |K^*|$-dimensional matrix $D = 0$. From this and from (A) it follows that G is a square nonsingular matrix (i.e., that $|S^*| = |K^*|$ and G^{-1} exists). Finally,

(C) states that strong complementarity holds; that is, the nonbasic coordinates of the primal solution have negative duals, whereas the constraints that vanish have positive dual variables.

THEOREM A6.2.2 (The local convergence theorem) Let the structure of the iterative aggregation process be characterized by the conditions (a)-(d) and let the sufficient conditions (A)-(C) for an isolated stable maximum be satisfied at the optimal point of the original problem (6.3.1), (6.3.3), (6.3.4). Further, suppose that the parameters α, γ, Q_0, Q satisfy one of the following conditions:

$$0 < \alpha < \frac{2}{q + J}, \qquad 0 < \gamma < 1; \tag{A6.2.41}$$

or

$$0 < \gamma < \frac{2}{(q + J)\alpha}, \qquad \alpha > 0, \tag{A6.2.42}$$

where $q = Q/Q_0$; $Q, Q_0 > 0$. Then

1. If the original problem (6.3.1)-(6.3.4) is a linear programming problem, the process converges locally in a neighborhood of an optimal point.
2. If the original problem is separable with respect to the blocks of the iterative aggregation process, then for any parameters α, γ there exists values of $\hat{q}$, $\hat{Q} > 0$ such that for any $0 < q < \hat{q}$, $Q > \hat{Q}$ satisfying (A6.2.41) or (A6.2.42) the process converges locally.
3. If the original problem is a convex programming problem, the previous result holds if it is also true that

$$0 < \gamma < \frac{1}{J - 1}. \tag{A6.2.43}$$

Several comments are appropriate about the conditions of the theorem. First, a problem separable with respect to the blocks of the algorithm is one where the original problem has an objective function $c(x)$ and constraint functions which can be written as the sum of J terms, where the jth term depends only on the vector x_j ($j = 1, \ldots, J$).

The difference in the local convergence of the linear and nonlinear problems can be more briefly (but not strictly) characterized as follows. To guarantee the convergence of the process to solve the linear problem it is sufficient to ensure that one of the constraints (A6.2.41) or (A6.2.42) on the parameters of the process is valid. In the nonlinear case it is also required that the term $q = Q/Q_0$ be sufficiently small while the parameter Q is sufficiently large.

The proof of the theorem is not given as it is very time-consuming and because the main ideas behind the proof are the same as the ideas in the proof given by Vakhutinsky (1979).

References

Abadie, J., and Williams, A. 1963. Dual and Parametric Methods in Decomposition. In *Recent Advances in Mathematical Programming*. McGraw-Hill Book Company, New York.

Ara, K. 1959. The Aggregation Problem in Input-Output Analysis. *Econometrica*, vol. 27, no. 2,

Arhangelsky, Yu. S., Vakhutinsky, I. Ya., Dudkin, L. M., Dyakonova, M. E., Novoselsky, I. A., and Khomyakov, V. A. 1975. Numerical Analysis of the Iterative Aggregation Methods for the Solution of the Interproduct Input-Output Problem. *Automation and Remote Control*, no. 7.

Arrow, K. J., Hurwicz, L., and Uzawa, H. 1958. *Studies in Linear and Nonlinear Programming*. Stanford University Press, Stanford, Calif.

Babadzhanyan, A. A. 1980. About the Convergence of General Processes of Iterative Aggregation. *Dokladii Akademii Nauk Armyanskoy SSR*, vol. 71, no. 5 (in Russian).

Bakhmetyev, M. M., Vakhutinsky, I. Ya., Dudkin, L. M., Livshitz, V. N., and Pozamantir, E. I. 1973. A Block Method of Computing the Optimal National Economy Plan. Proceedings of the 4th Winter School on Mathematical Programming and Adjacent Problems (Dragobich, 1971), vol. 3, CEMI Akademii Nauk SSSR, Moscow (in Russian).

Bakhmetyev, M. M., Vakhutinsky, I. Ya., Dudkin, L. M., Livshitz, V. N., and Pozamantir, E. I. 1978. Combined Method of Intercoordinating Systems of Planned Models of Transport and Production, in *Modelirovaniye Vnutrennich i Vneshnich Svyarey Otraslevich Sistem*. "Nauka" Publishing House (Siberian Branch), Novosibirsk (in Russian).

Belenyky, V. Z., and Volkonsky, V. A. 1974. *Iterative Methods in The Theory of Games and Mathematical Programming*. "Nauka" Publishing House, Moscow (in Russian).

Bell, E. J. 1965. Primal-Dual Decomposition Programming. Ph.D. thesis, Operation Research Center, University of California at Berkeley, Berkeley, Calif., Report ORS 65-23.

Brown, G. W. 1951. Iterative Solutions of Games by Fictious Play, in *Activity Analysis of Production and Allocation*, T. C. Koopmans (ed.). John Wiley & Sons, Inc., New York.

Chatelin, F., and Miranker, W. L. 1982. Acceleration of Successive Approximations Methods. *Linear Algebra and Its Applications*, vol. 43.

Chatelin, F., and Miranker, W. L. 1984. Aggregation-Disaggregation for Eigenvalue Problems. *SIAM Journal on Numerical Analysis*, vol. 21, no. 3.

Dantzig, G. 1963. *Linear Programming and Extension*. Princeton University Press, Princeton, N.J.

Dantzig, G., and Wolfe, Ph. 1960. Decomposition Principle for Linear Programs. *Operations Research*, vol. 8, no. 1.

Dantzig, G., and Wolfe, Ph. 1961. Decomposition Algorithm for Linear Programs. *Econometrica*, vol. 29, no. 4.

Dudkin, L. M. 1966. *An Optimal Input-Output Balance of a National Economy (Models for Current and Long-Term Planning)*. "Economica" Publishing House, Moscow (in Russian).

Dudkin, L. M. 1972. *A System of Calculations of Optimal National Economy Plan*. "Economica" Publishing House, Moscow (in Russian).

Dudkin, L. M., ed. 1979. *Iterative Aggregation and Its Application in Planning*. "Economica" Publishing House, Moscow (in Russian).

Dudkin, L. M., and Ershov, E. B. 1965. Input-Output Model and Material Balances of Individual Products. *Planovoe Khozyaistvo*, no. 5 (in Russian).

Dudkin, L. M., and Ivankov, S. A. 1975. An Economic and Geometrical Interpretation of the Process of Iterative Aggregation. *Economika i Matematicheskie Metodi*, vol. XI, no. 2 (English translation in *Matecon*, vol. XII, no. 3, 1976).

Dudkin, L. M., and Kasparson, V. A. 1979. Generalized Iterative Aggregation Processes for the Solution of the System of Equations of Interproduct Input-Output Model. *Automation and Remote Control*, no. 2.

Dudkin, L. M., and Kasparson, V. A. 1983. Transformation of Economic Iterative Aggregation Processes into Provably Convergent Processes. *Annals of the New York Academy of Sciences*, 0077-8923/83/0410-0201.

Dudkin, L. M., and Rabinovich, I. N. 1976. Iterative Aggregation for the Natural Interproduct Input-Output Model. *Economika i Matematicheskie Metodi*, vol. XII, no. 4 (in Russian).

Dudkin, L. M., and Vakhutinsky, I. Ya. 1975. Intercoordinating Optimal Plans of Different Levels of Management by the Help of Iterative Aggregation Methods, in *Multilevel Systems of Industrial Optimization*. "Nauka" Publishing House (Siberian Branch), Novosibirsk (in Russian).

Dudkin, L. M., Khomyakov, V. A., and Shchennikov, B. A. 1973. Basic Problems of Classical Aggregation in Input-Output Model. *Economica i Matematicheskie Metodi*, vol. IX, no. 2 (in Russian).

Dyakonova, M. E., and Novoselsky, I. A. 1979. Numerical Analysis of the Iterative Aggregation Process for Solving Optimal Interproduct Input-Output Model, in *Matematicheskoye Modelirovaniye Economicheskich System*. "Shtiinza" Publishing House, Khishenev (in Russian).

Dyumin, N. V., and Arhangelsky, Yu. S. 1966. Aggregation in Input-Output Analysis. *Economica i Matematicheskie Metodi*, vol. II, no. 6 (in Russian).

Ennuste, Yu. A. 1976. *Principles of Decomposition Analysis of Optimal Planning*. "Valgas" Publishing House, Tallin (in Russian).

Fedorenko, R. P. 1978. *Approximate Solution of Optimal Control Problems*. "Nauka" Publishing House, Moscow (in Russian).

Fiacco, A. V., and McCormick, G. P. 1968. *Nonlinear Programming: Sequential Unconstrained Minimization Techniques*. John Wiley & Sons, Inc., New York.

Fisher, F. M. 1965. Embodied Technical Change and the Existence of an Aggregate Capital Stock. *Review of Economic Studies*, vol. 32, no. 4 (92).

Fisher, F. M. 1968. Embodied Technology and the Existence of Labour and Output Aggregates. *Review of Economic Studies*, vol. 35, no. 4 (104).

Fisher, W. D. 1958. Criteria for Aggregation in Input-Output Analysis. *Review of Economics and Statistics*, vol. 40, no. 3.

Fletcher, R. 1973. An Ideal Penalty Function for Constrained Optimization, AERS Report CSS2.

Gale, D. 1960. *The Theory of Linear Economic Models*. McGraw-Hill Book Company, New York.

Gavriletz, Yu. N. 1963. Problem of Aggregation and Optimization of Economic-Mathematical Models, in *Economico-Matematicheskie Tetradi*, no. V, LEMM Akademii Nauk SSSR, Moscow (in Russian).

Gill, P. E., and Murray, W., eds. 1974. *Numerical Methods for Constrained Optimization*. National Physical Laboratory, Teddington, Middlesex. Academic Press, Inc., London.

Golshtein, E. G. 1966. About One Approach to the Solution of Linear Programming Problems with Block Structure. *Dokladi Akademii Nauk SSSR*, vol. 166, no. 5 (in Russian).

Golshtein, E. G. 1972. A Generalized Gradient Method for Searching of Saddle Points. *Economica i Matematicheskie Metodi*, vol. VIII, no. 4 (in Russian).

Golshtein, E. G., and Tretyakov, N. V. 1974. Modified Lagrangian Functions. *Economica i Matematicheskie Metodi*, vol. 10, no. 3 (in Russian).

Hatanaka, M. 1952. Note on Consolidation within a Leontief System. *Econometrica*, vol. 20, no. 2.

Hestenes, M. R. 1969. Multiplier and Gradient Methods. *Journal of Optimization Theory and Applications*, vol. 4, no. 5.

Kalinina, M. V. 1973. About the Convergence of an Aggregation Method for the Solution of Large-Scale Linear Equations Systems, in *Vichislitelniye Metodi i Programirovaniye*, no. 21. Moscow University Publishing House, Moscow (in Russian).

Kasparson, V. A. 1976a. About the Convergence of an Iterative Aggregation Process with Matrices Having a Little Norm. *Economica i Matematicheskie Metodi*, vol. VII, no. 1 (in Russian).

Kasparson, V. A. 1976b. A Proof of Convergence of a Modified One-Level Iterative Aggregation Process for the Matrices with Little Norm. *Automation and Remote Control*, no. 7.

Khisder, L. A. 1971. A Proof of Convergence of an Iterative Aggregation Process in a General Case, in *Issledovaniya po Matematicheskoy Economii i Smezhnim Voprosam*. Moscow University Publishing House, Moscow (in Russian).

Khomyakov, V. A. 1973. A Generalization of One Proof of Convergence of an Iterative Aggregation Process for the Solution of Linear Equations System, *Automation and Remote Control*, no. 7 (in Russian).

Kornai, J., and Liptak, T. 1965. Two-Level Planning. *Econometrica*, vol. 33.

Kossov, V. V. 1966. *Interindustry Balance*. "Economica" Publishing House, Moscow (in Russian).

Kuboniva, Masaaki. 1979. On the Optimization Approach to Socialist Planning and Pricing: Optimal Pricing, Labour Theory of Value and Planning with Aggregation of Goods. Institute of Economic Research, Hitotsubshi University, Discussion Paper Series, no. 19, March.

Kuboniva, Masaaki. 1984. Stepwise Aggregation for Material Balances. *Journal of Comparative Economics*, vol. 8.

Lasdon, L. 1970. *Optimization Theory for Large-Scale Systems*. Macmillan Publishers Ltd., London.

Leontief, W. 1947a. A Note on the Interrelations of Subsets of Independent Variables of a Continuous Function with Continuous First Derivatives. *Bulletin of the American Mathematical Society*, vol. 53, no. 4.

Leontief, W. 1947b. Introduction to a Theory of the Internal Structure of Functional Relationships. *Econometrica*, vol. 15, no. 4.

Levin, G. M., and Tanaev, V. S. 1974. Parametric Decomposition of Extremum Problems. *Izvestiya Akademii Nauk SSSR*, no. 4 (in Russian).

Lyubich, Yu. I., 1974. Iterations of Quadratic Representations, in *Matematicheskaya Economica i Functionalniy Analys*. "Nauka" Publishing House, Moscow (in Russian).

Malinvaud, E. 1954. Aggregation Problems in Input-Output Models, in *The Structural Interdependence of the Economy*. John Wiley & Sons, Inc., New York.

Malinvaud, E. 1968. *Procédures décentralisées pour la préparation des plans*. Institut National de la Statistique et des Etudes Economiques, Paris.

Manove, Michael, and Weitzman, Martin L. 1978. Aggregation for Material Balances. *Journal of Comparative Economics*, vol. 2.

Martinez Soler, F. 1973. A Process of Iterative Aggregation of Variables in Linear Optimization Planning Processes; Its Generalizations and Applications, in *Issledovaniya po Matematicheskoy Economii i Smezhnim Voprosam*. Moscow University Publishing House, Moscow (in Russian).

Martinez Soler, F. 1977. An Analysis of a General Iterative Aggregation Algorithm, in *Optimum Convex Models*. Moscow University Publishing House, Moscow (in Russian).

Martinez Soler, F., and Chernyak, V. I. 1974. *Modelling of Planning Calculations*. "Economica" Publishing House, Moscow (in Russian).

Mednitsky, V. G. 1972a. Aggregation in Linear Models. *Economica i Matematicheskie Metodi*, vol. VIII, no. 4 (in Russian).

Mednitsky, V. G. 1972b. About an Optimal Aggregation in a Block Linear Programming Problem, in *Matematicheskie Metodi Resheniya Economicheskich Zadach*, no. 3. "Nauka" Publishing House, Moscow (in Russian).

Mendelssohn, Roy. 1982. An Iterative Aggregation Procedure for Markov Decision Process. *Operations Research*, vol. 30, no. 1.

Miranker, W. L., and Pan, V. Ya. 1980. Methods of Aggregation. *Linear Algebra and Its Applications*, vol. 29.

Morimoto, Y. 1970. On Aggregation Problems in Input-Output Analysis. *Review of Economic Studies*, vol. 37, no. 1.

Movshovich, S. M. 1966. Method of Imbalances for the Solution of the Problems with Block Structure. *Economica i Matematicheskie Metodi*, vol. 2, no. 4 (in Russian).

Neudecker, H. 1970. Aggregation in Input-Output Analysis: An Extension of Fisher's Method. *Econometrica*, vol. 38, No. 6.

Polak, K. 1971. *Computational Methods in Optimization: A Unified Approach*. Academic Press, Inc., New York.

Polterovich, V. M. 1969. Block Methods of Convex Programming and Their Economic Interpretation. *Economica i Matematicheskie Metodi*, vol. V, no. 6 (in Russian).

Polterovich, V. M. 1970. *Mathematical Models of Reallocation of Resources*. CEMI Akademii Nauk SSSR, Moscow (in Russian).

Polyak, B. T., and Tretyakov, N. V. 1972. About One Iterative Linear Programming Method and Its Economic Interpretation. *Economica i Matematicheskie Metodi*, vol. VIII, no. 5 (in Russian).

Polyak, B. T., and Tretyakov, N. V. 1973. Penalty Estimates Method for the Conditional Extremum Problems. *Zhurnal Vychislitel'noi Matematiki i Matematicheskoi Fiziki*, vol. 13, no. 1 (in Russian).

Powell, M. J. D. 1969. A Method for Nonlinear Constraints in Minimization Problems, in *Optimization*, R. Fletcher (ed.). Academic Press, Inc., New York, Chap. 19.

Prilutsky, L. N. 1979. About Applications of Decomposition Algorithms, Which Use Modified Lagrange Functions, for the Coordination of Planned Calculations, in *Opit i Perspektivi Primeneniya Otraslevich ASU*. CNIITEI Priborostroyeniya, Moscow (in Russian).

Rabinovich, I. N. 1976. About the Convergence of an Iterative Aggregation Algorithm in the Optimal Interproduct Input-Output Model, in *Metodi Optimizatsii i Issledovaniya Operatsiy* (Prikladnaya Matematica). SEI SO Akademii Nauk SSSR, Irkutsk (in Russian).

Rasumikhin, B. S. 1967. An Iterative Method for the Solution and Decomposition of Linear Programming Problems. *Avtomatika i Telemekhanika*, no. 3 (in Russian).

Rasumikhin, B. S. 1976. *Physical Models and Equilibrium Theory Methods in Programming and Economics*. "Nauka" Publishing House, Moscow (in Russian).

Ritter, K. 1967. Decomposition Method for Programming Problems with Coupling Constraints and Variables. Mathematics Research Center, University of Wisconsin, Madison, Wis., Report 739.

Robinson, J. 1951. An Iterative Method of Solving a Game. *Annals of Mathematics*, vol. 54.

Rockafellar, R. T. 1970. *Convex Analysis*. Princeton University Press, Princeton, N.J.

Rockafellar, R. T. 1973. The Multiplier Method of Hestenes and Powell Applied to Convex Programming. *Journal of Optimization Theory and Applications*, vol. 12, no. 6.

Rockafellar, R. T. 1976. Monotone Operators and the Proximal Point Algorithm. *SIAM Journal on Control and Optimization*, vol. 14, no. 5.

Rosen, J. B. 1964. Primal Partition Programming for Block Diagonal Matrices. *Numerische Mathematik*, vol. 6.

Schwartz, B., and Tichatschke, R. 1975. Über eine Zerlegungsmethode zür Grossdimensionierter Linearer Optimierungaufgaben. *Matematische Operationsforschung und Statistik*, vol. 6, no. 1.

Shchennikov, B. A. 1965. Block Method for the Solution of Linear Equations System of Large-Scale Dimension. *Economica i Matematicheskie Metodi*, vol. 1, no. 6 (in Russian).

Shchennikov, B. A. 1966. Implementation of Iterative Aggregation Methods for the Solution of Linear Equations Systems. *Economica i Matematicheskie Metodi*, vol. II, no. 5 (in Russian).

Shor, I. Z. 1968. Implementation of the Generalized Gradient Descent in Block Programming. *Kibernetika*, no. 3 (in Russian).

Shpalensly, Yu. A. 1981. Two-Level Iterative Aggregation Algorithm for Unconditional Optimization Problem. *Automation and Remote Control*, no. 1.

Silverman, G. 1972. Primal Decomposition of Mathematical Programs by Resource Allocation, *Operations Research*, vol. 20, no. 1.

Telle, V. 1975. An Application of the Modified Lagrange Function in Block Programming. *Economica i Matematicheskie Metodi*, vol. XI, no. 3 (in Russian).

Telle, V. 1977. Processes for Coordinating of Plans of Different Levels of National Economy and Its Mathematical Modelling. Dissertation. CEMI Akademii Nauk SSSR, Moscow (in Russian).

Theil, H. 1957. Linear Aggregation in Input-Output Analysis. *Econometrica*, vol. 25, no. 1.

Tretyakov, N. V. 1973. Penalty Estimates Method for the Convex Programming Problems. *Economica i Matematicheskie Metodi*, vol. IV, no. 3 (in Russian).

Vakhutinsky, I. Ya. 1979. One Iterative Aggregation Process for a Convex Partially Separable Programming Problem. *Doklady Akademii Nauk SSSR*, vol. 195, no. 6 (in Russian).

Vakhutinsky, I. Ya. 1983. About a General Scheme for Coordinating Solutions in Two-Level Management System, in *Teoriya i Praktika Ispolsovaniya Metodov Agregirovaniya v Planirovanii i Upravlenii.* Materiali Soveshchaniya (Kazan, 1982), Akademiya Nauk Armyanskoy SSR, Erevan (in Russian).

Vakhutinsky, I. Ya., and Dudkin, L. M. 1973. Iterative Aggregation Algorithms for the Solution of the Linear Programming Problem of General Kind. *Izvestija Sibirskogo Otdelenija Akademii Nauk SSSR, Serija Obshchestvennich Nauk*, no. 11 (in Russian).

Vakhutinsky, I. Ya., and Kasparson, V. A. 1977. A Proof of Convergence of Modified Iterative Aggregation Processes for the Matrices with Little Norm. *Economica i Matematicheskie Metodi*, vol. XIII, no. 4.

Vakhutinsky, I. Ya., Dudkin, L. M., and Khomyakov, V. A. 1974. Two Approaches to the Problem of Aggregation (Classical and Iterative Aggregation). *Metron*, vol. XXXII, nos. 1-4.

Vakhutinsky, I. Ya., Dudkin, L. M., and Ryvkin, A. A. 1979. Iterative Aggregation: A New Approach to the Solution of Large-Scale Problems. *Econometrica*, vol. 47, no. 4.

Vakhutinsky, I. Ya., Dudkin, L. M., and Shchennikov, B. A. 1972. Iterative Aggregation Methods, in *Modelling of Economic Processes*. Moscow University Publishing House, Moscow (in Russian).

Vakhutinsky, I. Ya., Dudkin, L. M., and Shchennikov, B. A. 1973. Iterative Aggregation in Some Optimal Economic Models. *Economica i Matematicheskie Metodi*, vol. IX, no. 3 (translated into English in *Matecon*, vol. XII, no. 3, Spring 1976).

Vakhutinsky, I. Ya., Grigorova (Kalashnikova), G. A., and Dudkin, L. M. 1974. One-Level Iterative Aggregation Algorithms for the Intercoordination of Optimal Planning Problems of Some Industry, in *Problemi Optimizatsii Perspectivnogo Planirovaniya v Electrotechnicheskoy Promishlennosti*. Informelectro, Moscow (in Russian).

Vakhutinsky, I. Ya., Dudkin, L. M., Grigorova (Kalashnikova), G. A., Rabinovich, I. N., and Ryabova, E. A. 1975. One-Level Iterative Aggregation Algorithms. Proceedings of the 4th Winter School on Mathematical Programming and Adjacent Problems (Dragobich, 1973). CEMI Akademii Nauk SSSR, Moscow (in Russian).

Ven, V. L., and Erlich, A. I. 1970. Some Aggregation Question in Linear Models. *Izvestija Akademii Nauk SSSR, Serija Technicheskaja Kibernetika*, no. 5 (in Russian).

Volkonsky, V. A. 1965a. Optimal Planning in Large-Scale Conditions (Iterative Methods and Decomposition Principles). *Economica i Matematicheskie Metodi*, vol. 1, no. 2 (in Russian).

Volkonsky, V. A. 1965b. Current Planning Model and Approximation of Productive Opportunities of Some Oilgas Plant. In *Matematicheskie Metodi v Economike i Planirovanii*, Vol. 3, CEMI Akademii Nauk SSSR, Moscow (in Russian).

Waksman, V. S. 1969. About Matrices Which One Can Meet in Linear Economic Models. *Economica i Matematicheskie Metodi*, vol. V, no. 4 (in Russian).

Weitzman, M. 1970. Iterative Multilevel Planning with Production Targets. *Econometrica*, vol. 38, no. 1.

Yamada, I. 1961. *Theory and Application of Interindustry Analysis*. Tokyo.

Index